W9-CLN-060

THE QUANTUM AND THE LOTUS

A JOURNEY TO

THE FRONTIERS

WHERE

SCIENCE AND

BUDDHISM MEET

CROWN PUBLISHERS, NEW YORK

PREVIOUSLY PUBLISHED AS

L'INFINI DANS LA PAUME DE LA MAIN

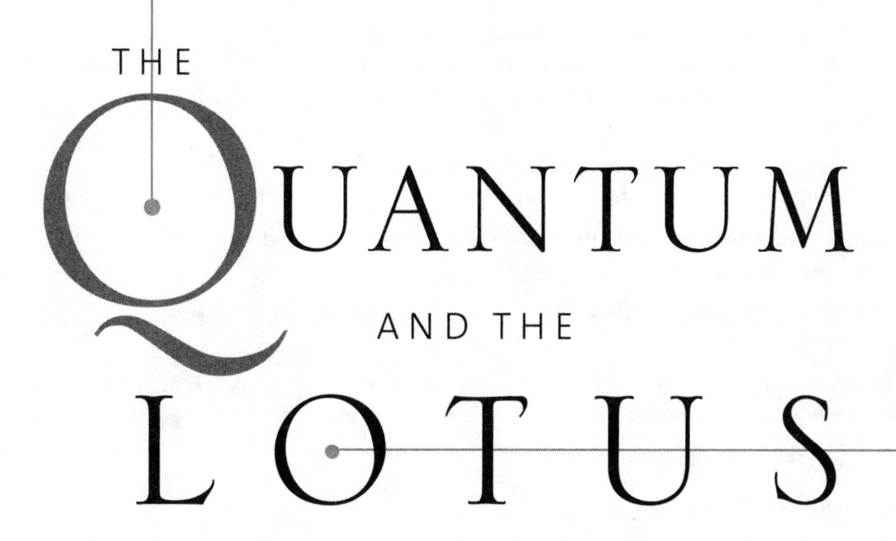

The Quantum and the Lotus

MATTHIEU RICARD AND

TRINH XUAN THUAN

Published by Crown Publishers, New York, New York.
Member of the Crown Publishing Group.

Random House, Inc. New York, Toronto, London, Sydney, Auckland
www.randomhouse.com

Printed in the United States of America

Design by Lauren Dong

Library of Congress Cataloging-in-Publication Data

Ricard, Matthieu.
[Infini dans la paume de la main. English]

The quantum and the lotus: a journey to the frontiers where science and Buddhism meet / Matthieu Ricard and Trinh Xuan Thuan—1st ed.
Includes bibliographical references and index.
1. Buddhism and science. 2. Physics—Philosophy. 3. Buddhism—Psychology. 4. Reality—Psychological aspects. 5. Physics—Religious aspects—Buddhism. 6. Spiritual life—Buddhism. 7. Enlightenment (Buddhism). I. Trinh, Xuan Thuan. II. Title.
BQ4570.S3 R5313 2001
294.3'375—dc21

2001028852

ISBN 0-609-60854-1

10 9 8 7 6 5 4 3 2 1

First American Edition

To our mothers

Contents

Introduction *By Matthieu Ricard*

How should I lead my life? How should I live in society? What is knowable? These three questions have puzzled mankind through the ages. Ideally, our lives should lead us to a feeling of plenitude, so that we have no regrets at the moment when we die. Life in society should inspire us with a sense of universal responsibility. Knowledge should teach us about both the nature of the world around us and about our own minds.

These same questions lie at the heart of the practices of science, philosophy, politics, art, social work, and spirituality. Artificially compartmentalizing these activities, as so often happens in our lives today, leads inevitably to diminished perspectives. Without a wisdom bred of altruism, science and politics are double-edged swords, ethics is blind, emotions run wild, and spirituality becomes illusory.

From the seventeenth century, the time of the scientific revolution, to the present day, many people have considered science to be synonymous with knowledge. The exponential increase in the accumulation of information driven by the rise of science is not about to slow down. Meanwhile, religious practice has declined in democratic, secular states, while often becoming more radical in religious states. The great spiritual traditions, whether they were dogmatic or based rather on pure contemplative experience, provided powerful ethical rules that people could use to structure and inspire their lives. As science has developed, many people have become disillusioned with the teachings of the world's religions, and a secular faith in the revelations of science and the efficiency of technology has evolved. Others, however, point out that science is incapable of revealing all truths, and that while technology has produced huge benefits, the ravages it has caused are at least as great. What is more, science is silent when it comes to providing wisdom about how we should live.

The correct view of science is as an instrument, intrinsically neither good nor bad. Either praising it to the skies or damning it is as senseless as applauding or criticizing strength. The strength of an arm can kill as well as save. Scientists are no better or worse than other human beings in general. Science does not produce wisdom. While the insights of science can help us change our world, only human thought and concern can enlighten us about the path we should follow in life. As a complement to science, therefore, we must also cultivate a "science of the mind," or what we can call spirituality. This spirituality is not a luxury but a necessity.

Over the last twenty years, a dialogue has been conducted between Buddhism and science, largely because of an interest in science shown by certain Buddhist thinkers, notably the Dalai Lama. In 1987, businessman Adam Engle and scientist Francisco Varela first organized what has become a regular series of encounters between the Dalai Lama and a number of eminent scientists: neurologists, biologists, psychiatrists, physicists, and philosophers. As the participants wrote: "An extraordinary quality of these meetings has been the open-minded yet critical attitude of the Buddhists and the scientists, both eager to expand their horizons by learning of the methods of inquiry and the insights of the other. Published accounts of these 'Mind and Life'[1] meetings have been received with growing enthusiasm by people interested in cross-cultural and interdisciplinary dialogue, especially concerning the nature of the mind."[2] Their titles include *Gentle Bridges; Healing Emotions; Sleeping, Dreaming and Dying;* and *Science and Compassion.*[3] They have been complemented by more-in-depth studies such as *The Embodied Mind,*[4] by Francisco Varela et al., and *Choosing Reality* and *The Taboo of Subjectivity*[5] by B. Alan Wallace. The conversations recorded in this book were undertaken in very much the same spirit.

The main difference between the pursuit of knowledge in science versus the same pursuit in Buddhism is their ultimate goals. In Buddhism, knowledge is acquired essentially for *therapeutic* purposes. The objective is to free ourselves from the suffering that is caused by our undue attachment to the apparent reality of the external world and by our servitude to our individual egos, which we imagine reside at the center of our being.

Buddhism stresses the importance of elucidating the nature of the mind through direct contemplative experience. Over the centuries it has devised a profound and rigorous approach to understanding mental states and the ultimate nature of mind. The mind is behind every experience in life. It is also what determines the way we see the world. It takes only the slightest change in our minds, in how we deal with mental states and perceive people and things, for "our" world to be turned completely upside-down.

Profound as the findings of Buddhism are, it is important to keep in mind as you read this book that the Buddha's teachings are not dogmatic. The teachings should be considered as the insights of a guidebook that allows the traveler to follow in the Buddha's footsteps. Buddhism stands ready to revise its beliefs at any moment if they are proved to be wrong. Not that it has any doubts about the basic truth of its discoveries, nor does it expect that the results it has built up over 2,500 years of contemplative science will suddenly be invalidated. The teachings of Buddhism are based entirely on experience. In its quest for knowledge, Buddhism does not run away from contradictions; it feeds on them. The countless metaphysical debates that it has conducted over the centuries with Hindu philosophers, and the dialogues that it continues to have with science and with religions, have allowed it to hone, focus, and widen its philosophical ideas, its logic, and its understanding of the world.

But Buddhism's open-minded attitude is not cheap opportunism. It has an impressive philosophical tradition to offer, as well as profound and inspiring texts on the contemplative life, and a spiritual practice that requires unbreakable perseverance. The inner transformation that leads to enlightenment is quite different from philosophical research or investigation in the natural sciences. Buddhism is basically a science of enlightenment.

The dialogues you are about to read are not intended to make science sound mystical, or to bolster the beliefs of Buddhism with the discoveries of science. The aim is to appreciate the way in which science fits into a larger conception of life that takes account of the important role of subjective experience. These dialogues seek to bridge the gap between the way things are and the way they appear. Buddhism can resolve the discomforting contradiction between the traditional view of objective reality, which considers that phenomena exist "out there" as really and solidly as they seem to do, and recent discoveries in science (physics specifically) that refute this sort of objective reality. One of the fathers of quantum physics, Werner Heisenberg, wrote: "I consider the ambition of overcoming opposites, including also a synthesis embracing both rational understanding and the mystical experience of unity, to be the *mythos,* spoken or unspoken, of our present day and age."[6]

These dialogues reflect the perspectives of two quite different lifestyles: that of an astrophysicist who was born a Buddhist, and who wanted to confront his scientific knowledge with his philosophical origins; and that of a Western scientist who became a Buddhist monk, whose personal experience has led him to compare these two approaches to reality.

Trinh Xuan Thuan's life encompasses three cultures: Vietnamese, French, and American. He was born in Hanoi in 1948 during the colonial war, six years before the French defeat at Dien Bien Phu. He was educated in French schools. Deeply impressed by French culture, in 1966 he decided to study physics in France, because he thought that it could answer some of the questions he was asking himself about the world. But General de Gaulle's famous speech made that very year in Phnom Penh, demanding the immediate retreat of all American troops from Southeast Asia, put an end to his plans. The Vietnamese government broke off diplomatic relations with France, and the Vietnamese people could no longer go there to study. After a year at the University of Lausanne, in Switzerland, Thuan left for the United States, where he was drawn to the California Institute of Technology, the Mecca of astrophysicists. In particular, Cal Tech ran a telescope with a mirror diameter of five meters on Mount Palomar; this was the largest telescope in the world in 1967. The shadow of Edwin Hubble, who discovered the galaxies outside the Milky Way and their outward motion, which led to the theory of the expanding universe, lay over the campus.

Thuan's studies coincided with an extremely exciting period in astrophysics, when many new phenomena were being discovered. As he puts it, "In such an intellectual ferment, I just had to become an astrophysicist." Since then, he has continued observing the universe and has become an expert in the formation of galaxies. He has also published a number of widely appreciated books on popular science.[7] He now teaches at the University of Virginia.

As for me, I studied to become a scientist. I did several years of research at the Pasteur Institute, in the cellular genetics department run by Professor François Jacob, who won the Nobel prize for medicine. It was a highly stimulating intellectual atmosphere. Over time, I became interested in the teachings of Buddhism. In 1967 I went to India to meet great Tibetan teachers. I became the disciple of one of them, Kangyur Rinpoche. For several years I returned each summer to the inspiring atmosphere of his hermitage/monastery in Darjeeling, while still pursuing my scientific research. In 1972, when I had completed my doctoral dissertation, I decided that, rather than take up a life of science, I would move to the Himalayas. I lived in India, then in Bhutan and Nepal, where I spent twelve years with my second teacher, Dilgo Khyentse Rinpoche. I had the chance to go to Tibet with him on several occasions, despite the tragic situation that still prevailed there after the Chinese invasion, and that still prevails. I now live in Shechen Monastery, near Kathmandu.

I first met Thuan during the Summer University in Andorra in 1997, and we had many fascinating discussions during our long walks together in the inspiring mountain scenery of the Pyrenees. This book was born from those friendly exchanges, which sometimes united us and sometimes divided us.

AT THE CROSSROADS 1

Are there any grounds for a dialogue between science and Buddhism? To find out, we must first clarify the pursuits of each, and then explore whether Buddhism (and spirituality in general) can complement science in important ways, particularly concerning ethics, personal transformation, knowledge of ourselves, and genuine spiritual insight. Buddhism has always been interested in questions that are also basic to modern physics. Might science therefore also help Buddhism in its exploration of reality?

MATTHIEU: You've made an impressive journey from Vietnam to your life as an astrophysicist in the United States. What drew you into a life in science?

THUAN: The 1960s were a golden age for astrophysics, with many great discoveries being made, such as the detection of the cosmic background radiation, which is the residual heat from the Big Bang; the discovery of pulsars, which are stars made entirely of neutrons; and the identification of quasars, which are celestial objects in far distant space, near the edge of the universe, that emit an extraordinary amount of energy. When I arrived in the United States, satellite exploration of the solar system had gotten into full swing. I can still remember the wonder of watching the first images of the surface of Mars transmitted by the space probe Mariner on a screen in our classroom. Those pictures of a dry, sterile desert told humanity that there was no intelligent life on Mars.

The canals that nineteenth-century astronomers thought they'd seen turned out to be nothing but optical illusions created by sandstorms. In the midst of such intellectual ferment, I just had to become an astrophysicist. Ever since, I've continued to marvel at the wonders of the universe, observing it through state-of-the-art telescopes, all the while thinking about its nature, origin, evolution, and destiny.

What did you find unfulfilling about your scientific career? Leaving a biology lab in Paris for a Tibetan monastery in Nepal is an unusual move, to say the least.

MATTHIEU: It was a natural progression, a step in an increasingly thrilling exploration of the meaning of life. All I did was leap from one stone to the next, go from one valley into another, into ever more beautiful realms. I followed where my passions led, while trying not to waste a single moment of this priceless human existence. I was lucky enough to live for thirty years alongside some remarkable Buddhist masters. This was a simple, direct experience, but also so profound that I always find it difficult to describe. You can recognize human and spiritual perfection when you see it, but the usual words that come to mind—wisdom, knowledge, goodness, nobility, simplicity, rigor, integrity—just aren't enough.

I think what everyone should be doing, before it's too late, is committing themselves to what they really want to do with their lives. Scientific research was interesting, of course, but I felt as though I was just adding a tiny dot of color to a pointillist canvas without knowing what the final composition would be like. So was it worth giving up all the unique opportunities of a human existence for that? In Buddhism, on the other hand, the point of departure, the goal to be reached, the means to that end, and the obstacles in the way are all perfectly clear. All you have to do is to look into your own mind and see that it is so often dominated by egoism, and that egoism derives from a deep ignorance of the true nature of ourselves and of the world. This state of affairs inevitably makes us and others suffer. Our most urgent task is to put a stop to this. The means to this end is to develop love and compassion, and to eradicate ignorance by following the path of enlightenment. As the days and years go by a tangible change takes place that creates a rare joy, exempt from hope or fear, which has constantly nourished my enthusiasm.

T: So why this conversation with a scientist?

M: One of Buddhist philosophy's main tasks is to study the nature of reality, and science offers many compelling insights into the nature of our world.

T: My work constantly raises questions about reality, matter, time, and space. Whenever I come up against such concepts, I can't help wondering how Buddhism deals with them, and how the scientific view of reality corresponds to the idea of reality in Buddhism. Do these two points of view coincide, are they opposed, or do they simply have nothing in common? I haven't studied Buddhist texts, so I don't have the knowledge necessary to answer such questions.

M: Is there a solid reality behind appearances? What is the origin of the world of phenomena, the world that we see as "real" all around us? What is the relationship between the animate and the inanimate, between the subject and the object? Do time, space, and the laws of nature really exist? Buddhist philosophers have been studying these questions for the last 2,500 years. Buddhist literature abounds with logical treatises, theories of perception, analyses of different levels of the world of phenomena, and psychological treatises exploring aspects of consciousness and the ultimate nature of our minds.

T: Are you saying that Buddhism is a science of the mind? Is it a science in the same sense as a natural science—that is to say, based on observation, with mathematics as its language?

M: The authenticity of a science doesn't necessarily depend on physical measurements and complex equations. A hypothesis can be checked by inner experience in a rigorous way. The Buddhist method begins with analysis and then often uses "thought experiments," which are hypothetical experiments conducted in the mind, but which lead to irrefutable conclusions, even though the experiments cannot be physically carried out. This technique is widely used in science.

T: That's right. Thought experiments are extremely useful in physics in particular. Einstein and other great physicists have used them not only to demonstrate physical principles, but also to point out paradoxes in some physical situations. For example, when studying the nature of time and space, Einstein imagined himself astride a particle of light. When thinking about gravity, he saw himself in an elevator falling through a vacuum.

I understand that the questions explored by modern physics echo the investigations of Buddhism in unexpected ways. But why is Buddhism interested in modern science, and in particular in physics and astrophysics?

M: Of course, modern science isn't Buddhism's main preoccupation. But there is interest in the findings of science because Buddhism has long been asking similar questions. Can separate, indivisible particles be the "building blocks" of the world? Do they really exist, or are they just concepts that help us understand reality? Are the laws of physics immutable, and do they have an intrinsic existence, like Platonic ideals? While not exaggerating superficial similarities, a study of both the differences and points of agreement between science and Buddhism may help us to deepen our understanding of the world.

Buddhist research is, above all, based on insights perceived through direct life experience, and is not bound by rigid dogma. It is ready to accept any vision of reality that is perceived as authentic. One of its main goals is precisely to bridge the gap between the way things really are and the way they seem to be. The Buddha often put his disciples on their guard against the dangers of blind faith. He said, "Investigate the validity of my teachings as you would examine the purity of gold, rubbing it against a stone, hammering it, melting it. Do not accept my words simply out of respect for me. Accept them when you see that they are true."

But the simple accumulation of knowledge is not enough. My teacher Khyentse Rinpoche said, "If you amass intellectual learning just so that you will be influential and famous, your state of mind is no different from that of a beggar sponging off the rich. Such knowledge will bring no advantage either to yourself or to others. As the proverb goes: 'Much knowledge, much pride.' How can you be of help to others unless you subjugate the negative tendencies that are anchored in your very being? To think that you can is just a joke—like a penniless beggar inviting the whole village to a feast."[1] There are many signs of success in the contemplative life. But the most important is that after a few months or years, your egoism has lessened and your altruism has increased. If attachments, hatred, pride, and jealousy still remain as strong as before, then you have wasted your time, gone down a blind alley and fooled other people. In contrast, knowledge of natural science allows us to influence the world, either constructively or negatively, while having relatively little effect on ourselves. It is obvious that since scientific knowledge has no connection with goodness or altruism, it cannot create moral values. So we need a contemplative science, in which the mind itself investigates the mind, in order to dispel the fundamental delusions that generate so much suffering for ourselves and others.

T: My understanding is that the Buddha's teaching was essentially practical. He said that our main objective in life should be to improve ourselves rather than worrying about the origin of the universe or the nature of matter.

M: When someone asked the Buddha about the origin of the universe, then kept on asking him questions that had nothing to do with spiritual progress, he remained silent. Buddhism is essentially a path toward enlightenment. It establishes a natural ranking between different forms of knowledge, particularly between those that help in this objective and those that are of little use, no matter how interesting they may be.

T: What does Buddhism mean by "enlightenment"?

M: A state of supreme knowledge, combined with infinite compassion. Knowledge, in this case, does not mean merely the accumulation of data or a description of the world of phenomena down to the finest details.

Enlightenment is an understanding of both the relative mode of existence (the way in which things appear to us) and the ultimate mode of existence (the true nature of these same appearances). This includes our own minds as well as the external world. Such knowledge is the basic antidote to ignorance.

But by ignorance we do not mean a simple lack of information. Rather we mean a false vision of reality that makes us think that things we see around us are permanent and solid, or that our egos are real. This leads us to mistake fleeting pleasures or the alleviation of pain for lasting happiness. Such ignorance also makes us build our happiness on others' misery. We are drawn to what satisfies our ego, and are repulsed by what might harm it. Thus, little by little, we create ever greater mental confusion until we behave in a totally egocentric manner. Ignorance perpetuates itself, and our inner peace is destroyed. Buddhism's form of knowledge is the final antidote to suffering. In this sense, I must admit that knowing the brightness of stars or the distance between them may have a certain utility, but it cannot teach us how to become better people.

T: That's exactly why I've thought that Buddhism ignores knowledge that doesn't directly influence our spiritual and moral evolution and our daily behavior. How can knowing about the origin of the universe and its destiny or the nature of time and space help us reach Nirvana?

M: Another man asked the Buddha some questions about cosmology. In reply, he picked up a handful of leaves and asked, "Are there more leaves in my hands, or in the forest?" "There are more in the forest, of course," replied the man. The Buddha went on, "Well, the leaves in my hand represent the knowledge that leads to the end of suffering." In this way the Buddha showed that certain questions are superfluous. The world has limitless fields of study, as numerous as the leaves of the forest. But if what we want more than anything else is enlightenment, then it is better to concentrate entirely on that aim and gather together only the knowledge that is directly relevant to our quest.

But experience shows that it is necessary to understand correctly the *nature* of the exterior world and of the ego, or what we term "reality," if we want to eliminate ignorance. That is why the Buddha made this the central theme of his teaching. He also emphasized the difference between how we perceive phenomena and their true nature, as well as the evil effects of such confusion. Mistaking a rope for a snake in a dimly lighted forest causes ungrounded fear. But as soon as light is cast on the rope and its true nature is revealed, then fear fades away. Buddhist investigations show that the individual ego and the external phenomena of our world are not separated. The distinction between "self" and "others" is purely illusory. Buddhism calls the true state of reality "emptiness,"[2] or the absence of intrinsic existence. One of our greatest errors is to believe in a solid reality to what we perceive. This idea of a solid reality has dominated Western philosophical, religious, and scientific thought for over two thousand years.

T: That's right. Up through the nineteenth century, classical science argued that objects had an intrinsic existence governed by well-determined laws of cause and effect. But quantum mechanics, which was developed at the beginning of the twentieth century, seriously undermined the idea that the basic ingredients of matter have such a definite existence, and also raised doubts about whether the world was governed by strict rules of cause and effect. The Buddhist idea of emptiness seems to be in harmony with the quantum view of reality. But can you explain more about just what emptiness means?

M: When Buddhism states that emptiness is the ultimate nature of things, it means that the things we see around us, the phenomena of our world, lack any autonomous or permanent existence. But emptiness is not at all a void, or the absence of phenomena, as early Western commentators on Buddhism thought. Buddhism does not at all espouse any form of nihilism, or the belief in nothingness. Emptiness does not correspond to nonexistence. If you can't speak of real existence, you can't speak of nonexistence either. The Fundamental Treatise of the Perfection of Wisdom says, "Those who become fixed on emptiness are said to be incurable."[3] Why incurable? Because while a belief in the real existence of phenomena is dissipated by meditation on emptiness, if you get attached to emptiness itself, making it an object of your belief, you fall into nihilism. The same text therefore goes on, "Consequently the wise abide neither in being nor in nonbeing."

As part of the quest to understand this true state of reality we call emptiness, Buddhism seeks to understand the existence, or nonexistence, of so-called indivisible particles of matter. According to Buddhism, learning to understand the essential unreality of things, which modern science has helped to clarify, is an integral part of the spiritual way. Knowledge of our spirits and knowledge of the world are mutually enlightening and empowering. The ultimate aim of both is to dissipate suffering.

T: You make a point that raises an issue I have long found troubling about the world of science. As you know, when I was nineteen I went to Cal Tech. There I rubbed shoulders with the greatest scientific minds, Nobel laureates and members of the National Academy of Sciences. I naively thought that their brilliance and creativity made them superior beings in terms of life in general and human relationships in particular. I was bitterly disappointed. You can be a great scientist, a genius in your field, and yet remain a dreadful person in daily life. This disparity shocked me. I think that Buddhism, or other forms of spirituality, might be able to complement science by filling in the areas where it falls short, particularly when it comes to ethics.

The history of science is full of examples of great scientists who were distinctly less inspiring when it came to their personal relationships. A striking example is Newton, who, with the possible exception of Einstein, was the greatest physicist ever. He behaved in a despotic manner toward his colleagues in the Royal Society of London, wrongly accused Leibniz of robbing him of the invention of the calculus (while both had invented it independently), and he treated his rival, John Flamsteed, the Astronomer Royal, in a terrible way. Even worse, the German physicists Philipp Lenard and Johannes Stark, who both won the Nobel prize for physics, enthusiastically backed the Nazis and their anti-Semitism by proclaiming the superiority of "German science" over "Jewish science."

Occasionally, but only too rarely, scientific genius and a keen sense of morals and ethics come together in a single person. This was the case for Einstein, whom *Time* magazine has named as the most remarkable personality of the twentieth century. During the First World War, Einstein fearlessly stood up to the Kaiser by signing an antiwar petition. With the growth of Nazism in Germany, he became an ardent Zionist, while also raising the problem of Arab rights in the planning of a Jewish state. He then emigrated to the United States, where, despite being a convinced pacifist, he backed armed intervention against Hitler. Realizing that the Allies had to beat the Germans to the invention of the atomic bomb, he wrote to President Roosevelt, thus inspiring the Manhattan Project. After the destruction of Hiroshima and Nagasaki, Einstein ardently protested against the spread of nuclear weapons. He opposed McCarthyism and used his immense prestige to attack all forms of fanaticism and racism.

But there were also shady aspects to Einstein's private life. He was an indifferent father and sometimes an unfaithful husband. He divorced his first wife and neglected their handicapped daughter. There is a sort of fault line in his personal life, as he himself said: "For a man like me there comes a decisive turning point in life after which you gradually lose interest in the purely personal and ephemeral side of things and channel all your efforts into the task of understanding."

M: The important point here is not to condemn one scientist or praise another. What matters is the total lack of correlation between scientific genius and human values. This allows us to put science in its proper place. We can then see it in the larger perspective of life and ask ourselves about its true use.

Spirituality, which I see as a process of personal transformation, does not simply complement science. It is a fundamental human need. This is the real problem of the scientific world. Personal transformation is no easy matter, even for people who dedicate their entire energies to it. So if it is seen as having only a secondary importance, the chances of success are even slimmer. To leave spiritual transformation in the background as a sort of optional extra, when it ought to be a core part of one's existence, throws a shadow over the entire scientific enterprise. Its intentions are unclear, its means often not properly gauged, and its results ambivalent. Without a fundamentally positive and enlightened motivation, the exploration of the limits of the possible inevitably takes precedence over the examination of what is desirable or indispensable.

Some scientists think that their work consists entirely of exploring and discovering, and that they aren't responsible for the use their results are put to. Such a position is a mere illusion, willful blindness, or, at worst, just plain dishonesty. Knowledge gives power, and power requires a sense of responsibility and an idea that we are accountable for the direct or indirect consequences of our actions.

Scientific research is often, but not always, conducted with excellent intentions. It then falls into the hands of politicians, military men, and businessmen who put it to dubious use. No one can ignore the close relationship between science, power, and economics. However, few scientists ever raise doubts about researches whose "misuse" is easy to predict. It is often only after the fact that they have doubts—as was the case for the fathers of the atomic bomb. Others do not even bother to hide behind the supposed neutrality of basic research and openly collaborate in the production of bacteriological weapons and other means to inflict suffering.

T: It is inexcusable for any scientist to work knowingly on the development of instruments of death and mass destruction. During the Vietnam War, I was shocked to hear that several great American scientists, including some Nobel laureates, were members of the "Jason Division"—a committee set up by the Pentagon to advise the military in the development of new weapons. I found it revolting that these great minds would meet each month in order to come up with weapons that would kill as many people as possible.

M: Between 1936 and 1976, the Swedish government sterilized sixty thousand people who were considered to be "inferior." Between 1932 and 1972, four hundred American citizens in Alabama, all of them poor and black, were used as unwitting guinea pigs by the Public Health Service in order to study the long-term development of syphilis. The patients were promised free health care, plus other minor advantages (including five thousand dollars to cover their funerals), if they agreed to go in regularly for checkups. In fact, they received no treatment at all. This was known as the Tuskegee Study of Untreated Syphilis in the Negro Male, and was quite simply a study of the evolution of untreated syphilis, conducted by doctors and respectable scientists who then published their results in equally respectable medical journals. Twenty-eight patients died of the disease and one hundred of secondary complications, while forty wives and nineteen babies were contaminated.

The study was abruptly broken off when a journalist, Jean Heller, brought it to the attention of the general public. Not one of the members of the Health Service that carried out the study expressed the slightest regret. But these were not Nazi doctors. They were civil servants and researchers living in a free country. The victims were finally given some small compensation, but not one doctor was brought to trial. It was only in 1997 that President Clinton apologized in the name of the American people.

In 1978, Dr. Hisato Yoshimura received the highest Japanese award for his work on "the science of environmental adaptation." During the Second World War, Dr. Yoshimura was the director of Unit 731, which carried out experiments on Chinese and Allied prisoners. An example of his studies on environmental adaptation consisted in plunging them into ice-cold water, then hitting them with hammers to determine when their limbs began to freeze. Other experiments included handing out chocolate contaminated with anthrax bacilli to Chinese children, to see how quickly they died. These examples are exceptions—generally, science makes an immense effort to improve the human condition—but they do show that science has no inherent ethics.

T: I firmly believe that scientists should not remain indifferent to the consequences of their research. They must accept the responsibility, especially if military leaders, politicians, and businessmen use their results to wage war, strengthen their power, or earn more money by exploiting the poor and damaging the environment.

M: The arms trade is in fact one of the most exasperating examples of the hypocrisy of rich countries. Ninety-five percent of the weapons made in the world are produced by the five members of the United Nations Security Council! Another example of a total failure of ethics and responsibility.

The same goes for the waste of resources in wealthy countries. Six billion U.S. dollars would give a basic education to the entire planet. Every year, in Europe and America, $12 billion is spent on perfumes, while the world spends $400 billion on illegal drugs and $700 billion on arms.[4]

T: All the same, basic research cannot be blamed for these aberrations. No more can human intelligence. They are simply tools.

M: Indeed. The pernicious or futile use of research results merely reflects ethical weakness. But this is no excuse. Some applications of scientific research, such as genetics and atomic energy, may have whipped up public interest, but most people are not really concerned with ethics. This is a matter for special committees, whose conclusions have little impact on daily life. Political opportunism and, even more, the sacrosanct laws of the free market dominate.

Glaxo, the American pharmaceutical company, is an excellent example. It has threatened to sue the governments of South Africa and Thailand if they produce their own tri-therapy medication for treating AIDS at an affordable price. Glaxo apparently could deprive millions of patients of the chance to live a few years longer. This is a flagrant and scandalous rejection of altruism.* What is more, AIDS research has plenty of funding in wealthy countries. Letting poor countries produce their own treatments would not decrease the company's turnover by a single cent, given that the sick in Africa and Asia cannot afford to buy American products. In Nepal, where I live, an estimated 5 to 10 percent of the population has been infected, and *nobody* is receiving tri-therapy. The drugs aren't even being imported. The treatment, currently dominated by Glaxo, costs about five hundred dollars a month, while the average monthly wage is fifty dollars. I can easily imagine how disgusted honest scientists are by this sort of business strategy.

Another striking example is the total inability of governments to limit the discharge of toxic gases into the atmosphere, even though it is quite clear that our lives will be seriously affected. Only a worldwide movement, based on each person's determination, can eliminate this. It is perhaps in this context that a nondogmatic spiritual approach, such as Buddhism, could play an important role.

---

*Glaxo has recently agreed to make its AIDS drugs available to developing countries at a significantly reduced price.

T: How's that?

M: When I say "nondogmatic," I mean an approach that doesn't condemn progress and naively call for a return to an outmoded way of life, and yet doesn't blindly agree that progress, in terms of annual economic growth and technological developments, is indispensable for human happiness. If our aim is to be profoundly satisfied with our existence, then some things are essential and others can easily be dispensed with. Buddhism's way of looking at the world allows us to draw up a priority list covering our goals and activities, and thus take control of our lives. Its analysis of the mechanisms of happiness and suffering clearly shows the divergent results of egoism versus altruism.

T: But what does this have to do with ethics?

M: The basis of ethics is extremely simple. Nothing is intrinsically good or evil. Good and evil exist only in terms of the happiness or suffering they create in ourselves or in other people. If we adopt a truly altruistic attitude, so that we are deeply concerned with the well-being of others, then this becomes the surest guide for our judgment. In our daily lives, we will then be able to see far more easily which actions will bring about more happiness and will relieve more pain. This is direct experience, and not a moral theory or a set of predetermined rules. It means paying constant attention to our motives. The mind has been compared to a crystal that takes on the color of the place where it has been placed. It is neutral. Our intentions determine the true nature of our actions, no matter what their appearances might be.

The point is neither to condemn those who are driven by hatred, greed, pride, or jealousy, nor to tolerate such destructive emotions as if they were intrinsic parts of existence. Instead, they are treated as symptoms of a disease that can be cured, if we make the necessary effort. Buddhism's approach is in fact extremely pragmatic. Scientific research provides us with information, but brings about no spiritual growth or transformation. By contrast, the spiritual or contemplative approach *must* lead to a profound transformation in the way we perceive the world and act on it. It is not enough to *know,* as in quantum physics, that our consciousness can't be isolated from the rest of reality. We must understand by personal *experience* that it is a part of that global reality. The move from theoretical knowledge to direct experience is the key to ethical problems. When our ethics reflect our inner qualities and guide our behavior, then they are naturally expressed in our thoughts, words, and deeds. They thus inspire others.

T: So it's a matter of making theory and experience coincide.

M: Yes, this is what brings out the true value of experience. It isn't enough to discover scientifically that our own consciousness is intimately bound up with the whole of reality. Our minds must assimilate the implications of this discovery, and our lives must change accordingly. Practicing Buddhists know that when they perceive their own interdependence with the world, they are filled with an irresistible compassion toward every living being—a compassion that radically transforms their existence. Those who have met the Dalai Lama, for example, know that a few moments in his presence are more eloquent than a hundred speeches about love and compassion.

As for the Buddhist method of discovery and transformation, it is generally a gradual one. It begins with absorption and study, it proceeds with intellectual analysis, culminating when we integrate into our being, thanks to meditation, a new way of looking at things and of behaving. In this context, to meditate means becoming *familiar* with this new perception of the world. Comprehension leads to meditation, which is then expressed in actions. We thus pass directly from knowledge to inner accomplishment and finally to active ethics.

Our society produces few wise men. It sets up ethics committees made up of great thinkers. In the Tibetan society where I live, it would be inconceivable to include on such committees people who did not possess indisputable human qualities in every sense of the term. It would be hard to imagine spiritual masters who could excel in teaching spirituality while at the same time remaining selfish, irritable, vain, or bad fathers. No one would ever consult them.

T: In the West, committees of "wise men" are generally chosen according to professional criteria. Human qualities are less important. And yet it is obvious that true wisdom is a matter of the heart as much as of the mind. The spiritual approach could provide us with a guide for personal conduct. In my field we are faced with numerous ethical problems, which will become even more pronounced in the twenty-first century, such as nuclear proliferation, the destruction of the environment, cloning, genetic manipulation, and perhaps the selection of certain human characteristics. Should research be controlled? The answer requires a great deal of thought, for it is also necessary to protect the freedom to create and to discover. The imagination must be allowed to express itself freely, otherwise it will die. We have already seen the disastrous effects on science that totalitarian regimes can have, for example in China or in the former Soviet Union. The Lysenko affair in Russia is a case in point. Because he had the support of Stalin and the Communist Party, Trofim Lysenko managed to gag all opposition and, from 1932 to 1964, to impose the idea that genes do not exist, despite the lack of any experimental proof of that contention. He thus succeeded in setting back Soviet biology and genetics by several decades.

M: The communist Chinese have recently tried to justify their claim of suzerainty over Tibet by saying that the study of blood groups revealed connections between the Tibetans and the Chinese. It is a bit as if France would claim ownership of England by arguing that the blood cells of British people resemble those of the French.

T: Society must be aware that some forms of research can go awry. Genetic "tinkering" could give rise to a new eugenics, with the idea of preserving so-called superior races and of eliminating "deviant" or "inferior" individuals. William Shockley, who received the Nobel prize in physics for inventing the transistor, spent the last years of his life promoting a sterilization program based on IQ.

I think that scientists shouldn't get involved in certain research without first weighing the moral implications. How should such decisions be made? Like you, I believe they should be based on altruism and the universal responsibility Buddhism speaks of. Science should organize its research so that it does not harm others. Unfortunately, this is easier said than done. It is extremely difficult for scientists to gauge the true repercussions of their research. To take a well-known example, when Einstein discovered the equivalence of matter and energy while working on his special theory of relativity, he could never have imagined that it would lead to the atomic bomb and the extermination of the inhabitants of Hiroshima and Nagasaki.

M: Another example is the rather hysterical reaction to the cloning of the sheep Dolly. Genetics or atomic physics is not the problem; it is the use that is made of them. As Adlai Stevenson, a candidate for president of the United States, said in a speech in 1952, "Nature is neutral. Man has wrested from nature the power to make the world a desert or to make the deserts bloom. There is no evil in the atom—only in men's souls." Science can both protect life and invent the weapons that destroy it. The idea is not to muzzle research—that would be undesirable and probably impossible in any case—but to give greater emphasis to those human qualities that should inspire researchers and decision makers. All of this is also true of intelligence, wealth, and physical strength, beauty, or power. They are all intrinsically neutral tools that can be put to good or bad use. This is precisely why one of the most important elements in Buddhist practice is the development of altruistic behavior.

To be and not to be 2

DOES THE UNIVERSE HAVE A BEGINNING?

The notion of a beginning of the world is fundamental to all religions and also to science. The Big Bang theory, which states that the universe originated about 15 billion years ago, along with time and space, is the best scientific explanation for how the universe began. Buddhism's approach to the question is different. It argues that the idea of a primordial beginning is mistaken, and that our world is one of an endless series of worlds. Was the Big Bang truly a primordial explosion, or the start of one particular universe in an incalculable series of universes without end or beginning? Is the notion of a beginning of time and the universe perhaps fundamentally flawed?

THUAN: In our present state of knowledge, the Big Bang is the theory that best explains the origin of the universe. We think the universe was created about 15 billion years ago when an unimaginably small, dense and hot concentration of energy exploded, in the process also creating time and space.

Since then, the universe has continued to expand. This theory was formulated after the American astronomer Edwin Hubble observed in 1929 that the vast majority of galaxies were moving away from ours—the Milky Way—at high speed. Even stranger, the more distant the galaxies were, the faster they were moving. A galaxy ten times the distance away moves ten times more rapidly. Scientists deduced from this observation that all of the galaxies had taken exactly the same time to move from their starting point to their present position. Now, if we imagine a film of the galaxies moving apart in the universe and we run it backwards in our minds, we discover that as all the galaxies retraced their paths, they'd meet up at the same point in space at the same moment. This insight led to the idea of a big explosion, or Big Bang, that kicked off the expansion. For many, the Big Bang replaced the religious idea of creation.

When it was first conjectured, the idea of the Big Bang met a great deal of resistance. However, some scientists immediately took the idea seriously. In 1922 the Russian meteorologist and mathematician Alexander Friedmann used Einstein's General Relativity theory[1] to construct a model of an expanding universe. The Belgian cosmologist-priest Georges Lemaître independently did the same thing in 1927 and dubbed the infinitesimally small original state of the universe the "primordial atom." The Russian-American physicist George Gamow worked out in 1946 that during the first 300,000 years of the universe's existence, its temperature and density were so great that none of its current structures (galaxies, stars, life) could have existed and that it contained only elementary particles and radiation. According to Gamow, that hot primordial radiation should still be reaching us, though greatly cooled down by the energy lost during the 15 billion years of the universe's expansion.[2]

MATTHIEU: This is what's called the cosmic background radiation, right?

T: Yes, it's what's left of the heat of the "fire of creation," and its detection in 1965 offered our best proof yet for the theory. But no one bothered to look for the background radiation for many years, and it was discovered in 1965 only by accident. There were two reasons for this delay. First, the idea of the Big Bang was embraced by many theologians, which put some astrophysicists off. In 1951, Pope Pius XII likened God's words in the book of Genesis, "Let there be Light," to the explosion of the Big Bang.

The other reason scientists failed for so long to look for the background radiation is that a competing theory to that of the Big Bang was put forward that sidestepped the issue of creation. The Steady State theory was devised by three British astronomers, Hermann Bondi, Thomas Gold, and Fred Hoyle. This theory argued that the universe was in a "steady state," meaning that it had always been and would always be pretty much the same as it is now. In other words, it had neither a beginning nor an end. But observations soon undermined this theory. In the early 1960s, quasars were discovered. These are celestial objects at the edge of the universe that emit enormous energy from a very compact volume. Radio galaxies, which emit most of their energy as radio waves, were also discovered. Observations showed that the population of both quasars and radio galaxies was changing, that they became fewer in number with time, which contradicted the idea of an unchanging steady state.

Then, in 1965, with the unexpected detection of the cosmic background radiation, the Steady State theory received its final blow. The theory argued against the notion of a beginning and of a superdense and superheated explosion, and couldn't explain the presence of the remains of this primordial heat that still bathes the entire universe. So the Big Bang theory became the widely accepted explanation for the origin of the universe. The Big Bang was the only explanation that could account for such apparently unrelated phenomena as the outward motion of galaxies, the cosmic background radiation, and other important characteristics of the universe, such as the chemical makeup of stars.

M: But how could such a massive explosion result in the evolution of the universe? What immediately followed the Big Bang, and how, according to the theory, did the universe take shape?

T: Physicists say that the universe was born from a vacuum—they call it the quantum vacuum—but this vacuum was not calm and peaceful, as you might imagine. The quantum vacuum was seething with energy, even though it contained no matter. What seems to be empty space is filled with energy fields that can be described as waves. In fact, the space that surrounds us is filled with a veritable kaleidoscope of waves of different kinds.

Radio waves have the least energy. With a simple turn of a switch, modern electronics magically converts them into a Beethoven symphony or a TV show. Visible light rays, from the sun, constantly ricochet off the surface of the objects around us and then enter our eyes, allowing us to see. Solar ultraviolet rays are also present. Just as the space around us is teeming with such waves, the quantum void that existed at the birth of the universe was bursting with fields of energy.

In a remarkable process that is mind-boggling in its scale, this enormous energy drove the universe to expand extremely rapidly. Astrophysicists call this phase of the universe's evolution the period of "inflation." This inflation led to a staggering increase in the volume of the universe in practically no time at all. Between 10^{-35} and 10^{-32} seconds after the Big Bang, the universe grew exponentially from being much smaller than a hydrogen atom to the size of an orange.

As the universe expanded, it also grew much cooler. Right after the Big Bang, the universe was hotter than all of Dante's infernos put together, and this incredible heat prevented the formation of matter. As the universe cooled, energy began to be converted to matter according to Einstein's famous formula $E = mc^2$. As Einstein discovered, a quantity of energy can be converted into a particle of matter (and its mass, *m*, will be equal to the quantity of energy, *E*, divided by the speed of light, *c*, squared). From this point on, the history of the universe is a long ascension toward complexity.

Elementary particles (quarks and electrons, for instance) rose out of the primordial vacuum and came together to form atoms, then molecules, and finally the stars. Those stars assembled together to create galaxies, each containing several hundred billion stars, and the hundreds of billions of galaxies in the observable universe formed an immense tapestry covering the cosmos. The infinitely small had created the infinitely large. In at least one of these galaxies, the Milky Way, near a star called the Sun, on the planet Earth, molecules locked together to form long chains of DNA, which created life, then consciousness, and finally people capable of asking questions about the world around them and the universe that had caused them to exist.

M: No matter how convincing this theory is about the evolution of the universe, it doesn't explain the cause of the Big Bang. When I mentioned the Big Bang to a learned Tibetan friend, he exclaimed, "So the universe, time, and space all begin with a 'pop,' *ex nihilo,* without a cause? But that's as illogical as postulating the existence of a creator who is his own cause!"

According to Buddhism, time and space are just concepts created by our perception of the world, and have no existence apart from our perception. In other words, they are not "real." The idea of an absolute beginning of time and space is therefore flawed according to Buddhist thinking. We also believe that nothing, not even the apparent start of time and space, can come about without causes or conditions. In other words, nothing can start to exist or cease to exist. There can only be transformations. The Big Bang must then be a mere episode in a continuum without a beginning or an end.

T: You raise a troubling issue for the Big Bang theory. The truth is that we cannot say what happened "before" the Big Bang. I'm putting quotes around "before," because if time started with the Big Bang, then the idea of a "before" becomes meaningless.

Does science allow us to go all the way back to the moment of creation? The answer is no. Right now, there is a "wall" in the way, which is called Planck's wall, named after Max Planck, the German physicist who first studied the problem. This wall stands at the infinitely small time of 10^{-43} seconds (the number 1 preceded by forty-three zeros). This is called Planck time. At that moment the universe was ten trillion trillion times smaller than a hydrogen atom. Its diameter was equal to the "Planck length," or 10^{-33} centimeters.

M: Are you saying that this is some kind of natural limit? That we can never know what happened before this? Or is this barrier caused by our lack of knowledge?

T: Planck time isn't an absolute limit. Our inability to see what happened before this time is simply a product of our ignorance. At the moment we don't know how to unify the twentieth century's two great physical theories—quantum mechanics and relativity. The former describes the infinitely small, and accounts for the behavior of atoms and light when gravity isn't dominant. The latter describes the infinitely large, and allows us to understand the universe and its structures at a cosmic scale, where the two nuclear forces and the electromagnetic force are not predominant. And that's the snag. We don't know how to describe the behavior of matter and light when the four fundamental forces[3] are on equal footing, which is the way things were at Planck time, or 10^{-43} seconds after the Big Bang.[4]

Behind this artificial wall lies a reality that is still unknown to physicists. Some think that space and time, which are so intimately linked in the universe today, were separate in the early universe. According to this scenario, time no longer exists. The concepts of "past," "present," and "future" lose all meaning. Divorced from its partner, time, space becomes a quantum "foam" with constantly changing and shifting forms.

Some physicists, who are working on the theory of superstrings,[5] say that this quantum foam doesn't exist. According to their alternative theory, elementary particles are created by the vibration of tiny "strings" of energy that are Planck length long. Because nothing can be smaller than these strings, the problem of what happens to space on scales smaller than Planck length disappears. Space simply can't have a smaller dimension. This theory seems to have the potential of unifying quantum physics and relativity. But right now it's wrapped in a thick mathematical veil and hasn't been proved experimentally.[6] If we accept the existence of a quantum foam, then possibly one of these innumerable shifting forms in the quantum foam created the universe and space-time about 15 billion years ago. Before then, it's impossible to say that space spent so long in this or that form, since time didn't exist. An infinite period may even lie behind Planck's wall.

M: When you say "infinite period," do you mean without a beginning?

T: Anything is possible. We simply arrived at 10^{-43} seconds by extrapolating the known laws of physics back toward zero time. But they break down behind this wall. So it is that our current physics begins at 10^{-43} seconds after the Big Bang.

M: For Buddhism, the reality of our universe is seen from a quite different perspective. Buddhism considers that phenomena aren't really "born," in the sense that they pass from nonexistence into existence. They exist only in terms of what we call "relative truth," and have no actual reality. Relative, or conventional, truth comes from our experience of the world, from the usual way in which we perceive it—that is, by supposing that things exist objectively. Buddhism says that such perceptions are deceptive. Ultimately, phenomena have no intrinsic existence. This is the "absolute truth." In these terms, the question of creation becomes a false problem. The idea of creation becomes necessary only if we believe in an objective world.

The Buddhist view does not, however, exclude the possibility of the unfolding of the world. Obviously the phenomena we all see around us aren't nonexistent, but if we examine *how* they exist, then we soon see that they can't be viewed as a set of independent entities, each with its own existence. Thus, phenomena exist only as a dream, an illusion or mirage. Like mirror images, they can clearly be seen, but have no separate existence. Nagarjuna, the great second-century Indian philosopher, said, "The nature of phenomena is that of mutual dependence; in themselves, phenomena are nothing at all." Their evolution is neither random nor fixed by divine intervention. Instead, they follow the laws of cause and effect in a global interdependence and reciprocal causality. The problem of an "origin" comes about only from a belief in the absolute reality of phenomena and the existence of space and time.

In terms of absolute truth, there is no creation, no duration, and no end. This paradox is a good illustration of the illusory nature of the world of phenomena. It can reveal itself in an infinite number of ways because its final reality is emptiness. In terms of the relative truth of appearances, we say that the conditioned world, called samsara, is "without beginning" because each state must have been caused by the previous one. So, with the Big Bang theory, do we have an *ex nihilo* creation, a creation out of nothingness, or the expression of some kind of preexisting potential that is not yet manifested in the universe? Is it seen as a real beginning, or as a stage in the universe's evolution?

T: As we've just discussed, modern physics can't look into the time before Planck's wall. So, before the Big Bang, there could easily have been an infinite period of time, or else no time. Another possibility is that the universe is cyclical, and the Big Bang is just the start of one cycle among an infinite number of cycles. In either case, the issue of how the universe could possibly have arisen, *ex nihilo,* at zero time is avoided. Those possibilities are ways of dodging the creation problem. But they are still pure speculation, unsupported by observation or experiment.

M: Perhaps the Big Bang can be interpreted as the process of the world of phenomena springing forth from an infinite but nonmanifest potentiality, which is metaphorically called "particles of space" in Buddhism. This term doesn't refer to particles in the sense of bits of matter, but rather to space's potential. This potential could perhaps be compared to the vacuum of physics you described, so long as we don't invest the potential with any form of concrete or independent "reality."

In Buddhism, we believe there can be no *ex nihilo* creation. As Shantideva wrote in the seventh century:

When nonbeing prevails, if there's no being,
How could being ever supervene?
For insofar as entity does not occur,
Nonentity itself will not depart.

And if nonentity is not dispersed,
No chance is there for entity to manifest.
Being cannot change and turn into nonbeing,
For otherwise it has a double nature.[7]

The reason why "nothing" can't become "something" is that in order to do so, the "nothing" would be done away with. But how is it possible to get rid of something that does not exist? Nothingness is a mere concept defined in relation to existence. It does not have the slightest reality on its own, because it cannot be conceived in the absence of existence. Nothingness cannot be transformed. If something appears, it means that the potential for manifestation was already present.

T: Physics says that the potential for manifestation lies in the vacuum's energy. But we are still left with the question: How was the vacuum created? Was there nothing, then a sudden rupture, with the appearance of a vacuum full of energy, and simultaneously of time and space?

M: A causeless rupture, making nothing become something—that is quite a way to start! The Big Bang, or any other "beginning" of a given universe, can't happen without a cause and conditions. The world of phenomena can't have come from nowhere. One of Buddhism's essential ideas states that because things have no independent reality, they can't really"begin" or "end" as distinct entities. When we speak of a "beginning," our mind immediately pictures "something." The idea of the universe beginning and ending belongs to relative truth. In terms of absolute truth, it's meaningless. When you consider a castle seen in a dream, for instance, you don't need to worry about who actually built it. All religions and philosophies have come unstuck on the problem of creation. Science has gotten rid of it by removing God the Creator, who had become unnecessary. Buddhism has done so by eliminating the very idea of a beginning.

T: Do you remember the story about the great eighteenth-century French mathematician and physicist Pierre-Simon de Laplace? When he gave Napoleon a copy of his great book on celestial mechanics, the emperor scolded him for not once mentioning the "Great Architect." Laplace replied: "But, Your Highness, I have no need of that hypothesis." Questions still remain, however: Why is there a universe? Why are there laws? Why was there a Big Bang? We return to Leibniz's famous question: "Why is there something rather than nothing? For nothing is both simpler and easier than something. Moreover, assuming that things must exist, there must be a reason why they exist thus and not otherwise."

M: One reply would be the famous dictum of the second-century master Nagarjuna: "Since all is empty, all is possible." And the famous scripture Perfection of Wisdom says specifically, "Though phenomena appear, they are empty; though empty, they appear." In Buddhism, emptiness isn't just the true nature of phenomena, it's also the potential that allows the propagation of an infinite variety of phenomena. To use a simple metaphor, the continents, trees, and forests can exist because space allows them to. If the sky were made of rock, nothing much would happen. In the same way, if reality were permanent, and its properties too, then nothing would change. Phenomena could not appear. But because things have no intrinsic reality, they can have infinite manifestations.

When you have understood that everything is intrinsically empty, it's easier to understand how things work in relative, or conventional, truth. Even though phenomena lack reality, they don't happen at random. This is the emptiness of Buddhism. It isn't nothingness, but rather the absence of any permanent and autonomously existing phenomena.

T: Yes, but many people associate emptiness with nothingness. In the nineteenth century, Buddhism was accused of being nihilistic.

M: That was a serious mistake. We consider that there are two opposing and erroneous points of view: nihilism and materialistic realism. The latter, which Buddhism calls "eternalism," reifies the world by postulating the existence of immutable matter made of solid parts. What is more, when Leibniz wondered why there is "something rather than nothing," he presupposed that there really is *something.* In Buddhism's Middle Way, there is neither nothing (nihilism) nor something (materialism or realism). We could now ask Leibniz, "Why should there be nothing, since phenomena are possible?" The true nature of interdependent phenomena goes against common sense because these phenomena can't be called either existent or nonexistent. The intellect has its limitations, and we can't grasp the true nature of reality just by means of ordinary conceptual processes. Only direct knowledge that transcends conventional thought can see the world of phenomena in a nondual way, in which subject and object have become meaningless.

T: What, then, is the Buddhist explanation of how our world came about? Does Buddhism have a cosmology?

M: Yes, it does. But in a nondogmatic way. While some parts of the description have now become outdated, they continue to have a symbolic meaning in meditation. These parts of the cosmology conformed to the ideas of the time, which Buddhism adopted but which were not taught by the Buddha himself as absolute truths. This cosmology in no way contradicts Buddhism's analysis of "reality." It is important to bear in mind that the idea of the universe coming together belongs to relative truth and the world of appearances. So, in relative terms, Buddhism talks of the formation of our universe out of "particles of space" that don't stand for bits of matter, as I've said, but rather for potentials. Out of these particles of space a "void" formed that was "full" of five "winds" or energies (*prana* in Sanskrit). These energies appeared as lights of five different colors that gradually materialized into five elements—air, water, earth, fire, and space. When combined, they formed a "soup," an ocean of elements that was whipped up by the primordial energy, thus producing the heavenly bodies, continents, mountains, and finally living beings. This is the formation of one universe among the infinite number that exist. There is no idea of initial creation, because no one prime cause can be accepted.

T: Despite the imagistic language, this description of the world's beginning resonates well with modern scientific ideas. From both perspectives we are a million miles from seeing the birth of the universe as the result of the gods' loves and rivalries. I'm particularly intrigued by the notion of the "full void" we find in both science and Buddhism.

M: Yes, but be careful. There's a fundamental difference. Science talks of the universe as a truly existing, independent object. Buddhism, as we shall see, considers that the universe is not independent from consciousness. We could say that subject and object mold each other; the object does not exist without the subject, and vice versa.

Buddhism envisages a cyclical universe (though neither circular nor repetitive, as it was for the Stoics). Each cycle is made up of four periods: formation, continuity, destruction, and the "nonmanifest" state, which is the intermediate void between two universes. Particles of space ensure that there is continuity between one universe and the next. This cycle has neither an end nor a beginning.

T: Modern cosmology also contains the idea of a cyclical universe. If there is enough matter in our universe, then the force of gravity will, at a certain time, stop the universal expansion and pull the galaxies the other way. We will then have a Big Crunch, the opposite of a Big Bang. The stars will evaporate in the intense heat, and matter will fall apart into elementary particles. The universe will end its life in a blinding implosion into an extremely small, hot, dense state. Time and space will once more become meaningless. Can a universe that has collapsed into itself be reborn, phoenix-like, from its ashes with, perhaps, different physical laws? No one knows. For, as we were just saying, modern physics loses its grip at Planck time, and can't deal with such extreme temperatures and densities.

M: According to the Buddhist view, the end of a cycle is marked by a final conflagration. It is said that seven increasingly intense fires will burn up the visible world. The universe is then absorbed once more into the void, and a new cycle can begin.

T: According to modern cosmology, if the universe did start other cycles, then they would all be different. The universe would build up more and more energy, so that each new cycle would last longer than the previous one and the universe's maximum size would get bigger and bigger. It's rather like the idea of nonrepetitive cycles you just mentioned. But if our universe doesn't contain enough matter for gravity to stop its expansion, it will continue spreading until the end of time and so won't be cyclical. In the end, the stars will burn up all their nuclear fuel and go out. They will no longer light up the heavens. The world will be plunged into a long, dark night. Heat will fade away and the temperature get closer and closer to absolute zero, but without ever reaching it. With no more energy to keep it going, life as we know it will disappear. In the far distant future, the universe will be an immense sea of radiation and elementary particles.

According to our present state of knowledge, if the universe's expansion is to be reversed, then the universe must contain much more matter than has so far been observed. That said, finding out how much matter the universe contains is not an easy job given the large amount of "dark matter" (at least ninety percent of the total) that doesn't send out any radiation and just has a gravitational effect on its surroundings. Without light, astronomers are quite literally in the dark.

Recent observations of supernovas (exploding stars) in distant galaxies seem to show that the expansion of the universe, rather than slowing down, which would be the case if gravity were the only factor in question, is actually accelerating. This accelerating expansion implies the existence of a mysterious "dark energy," also called "quintessence." If this turns out to be right, then the universe will expand forever. There could be no new Big Bang to start a new cycle without there first being a Big Crunch, which would be impossible if the universe truly were expanding more rapidly all the time. But the properties of distant supernovas have not been properly understood yet, and this conclusion requires further verification. I should also mention that recent preliminary observations of temperature fluctuations of the cosmic background radiation also appear to indicate the presence of that "dark energy."

M: We aren't just talking about matter when we depict the universe as a series of metamorphoses without a beginning. Consciousness has no start, either. In Buddhism, the matter/consciousness duality, the so-called mind-body problem, is a false problem, given that neither of them has an intrinsic, independent existence. The problem of the lack of any first cause for phenomena and consciousness falls into the category of what Buddhism calls the "inconceivable." But we need not stand in dumb incomprehension in front of such mysteries unsolvable by the intellect. Certain things simply cannot be grasped using ordinary concepts. The idea of a beginning is "inconceivable" not because it would be so long ago or far off in space, but because our discursive minds cannot stand back from that process of beginning in a way that is needed to transcend all concepts. Our ordinary way of thinking emerged from that same process and thus it can't place itself "outside" the chain of causes and so determine its own origin.

T: That argument sounds remarkably like the scientific argument about insolvability known as Gödel's incompleteness theorem. This theorem, devised by the famous mathematician Kurt Gödel, implies that we are always limited in our knowledge of any system, such as the system we call the universe, while we are ourselves a part of that system. In any case, modern science has nothing to say about the possible coexistence of consciousness and matter since the beginning of time.

M: The notion of "inconceivability" helps to drive away the reifying instinct that makes us see phenomena as autonomous "real" entities. Since the ultimate nature of phenomena lies beyond the intellectual concepts of "existence" and "nonexistence," it can be called "inconceivable" without this being synonymous with ignorance. On the contrary, pure awareness, the non-dual knowledge of the ultimate nature of both mind and phenomena, beyond all concepts, is a quality of enlightenment.

According to some Buddhist teachings (the *tantras*), which analyze the manifestations of phenomena at a more contemplative level, the primordial nature of phenomena lies above and beyond notions of subject and object, or time and space. But when the world of phenomena emerges from primordial nature, we lose sight of this unity and make a false distinction between consciousness and the world. This separation between the self and the nonself then becomes fixed, and the world of ignorance (or *samsara*) is born. The birth of *samsara* did not happen at any particular time. It simply reflects at each instant, and for each of our thoughts, how ignorance reifies the world.

In Search of the Great Watchmaker

3

IS THERE A PRINCIPLE OF ORGANIZATION?

When we consider the remarkable harmony and precision of the universe, it is tempting to imagine that there is an all-knowing Creator or, from the secular view, some sort of principle of creation that finely adjusted the evolution of the universe. The omnipotence of such a Creator would explain everything, and there would be no need for us to wonder about the origins of our astonishingly complex universe, or about how life arose, or how inanimate matter can be compatible with the animate matter of life. This question of whether or not there is a creating God is a key point of distinction between the world's great spiritual traditions. For Buddhism, the notion of "first cause" does not stand up to analysis. Some scientists also dismiss the need for a God, arguing that the exceptional fine-tuning of the universe arose by chance. Others, however, believe that there is some kind of an organizing principle at work in our world. Can this notion of such a principle stand up to analysis? Is it necessary and logical?

THUAN: Since the sixteenth century, the place of humanity in the universe has gotten smaller and smaller. In 1543, Nicolaus Copernicus, a Polish priest, knocked the Earth off its pedestal as the center of the universe and discovered it was just another planet revolving around the Sun. Ever since, the ghost of Copernicus has continued to haunt us. If our planet wasn't at the center of the universe, then, our ancestors thought, the Sun must be. But along came an American astronomer, Harlow Shapley, who discovered that our sun is just a suburban star among the hundreds of billions of stars that make up our galaxy. We now know that the Milky Way is only one of the hundred billion or so galaxies in the observable universe, which has a radius of about 15 billion light-years. Humanity is just a grain of sand on this vast cosmic beach.

Human life has also become smaller in terms of time. Now we know that on a timeline that shows the 15 billion years of the universe as one year, the first human appears only at 10:30 P.M. on December 31 (about 3 million years ago). Stonehenge is built and Egyptian civilization arises at 11:50:54 P.M. (about 3,000 years ago). The Buddha appears on the timeline at 11:59:55 P.M. (2,500 years ago), and Christ shows up at 11:59:56 P.M. (2,000 years ago). The European Renaissance occurs at 11:59:59 P.M. (450 years ago), on the last day of the year.

This shrinking of our place in the world led to Pascal's cry of despair in the seventeenth century: "The eternal silence of endless space terrifies me."[1] Pascal's words were echoed three centuries later by the French biologist Jacques Monod: "Man knows at last that he is alone in the unfeeling immensity of the universe, out of which he has emerged only by chance." And the American physicist Steven Weinberg remarked, "The more the universe seems comprehensible, the more it also seems pointless."[2]

Personally, I don't think that human life emerged purely by chance in an unfeeling universe. To my mind, if the universe is so large, then it evolved that way in order to allow us to be here.

MATTHIEU: Careful, that sounds like what Bernardin de Saint-Pierre said: "Pumpkins have slices because they're made for family eating!"

T: We must indeed be careful about arguments based on justifications of final causes. Science was itself born from a total and categorical rejection of any such teleological thinking, which is the province of religious doctrines. That said, modern cosmology has discovered that the conditions that allow for human life seem to be coded into the properties of each atom, star, and galaxy in our universe and in all of the physical laws that govern it.

The way our universe evolved depended on what are called "initial conditions" and on about fifteen numbers called "physical constants." Newton's law of gravity depends on one of these constants, a number called the "gravitational constant," which determines the strength of gravity's attraction. In the same way, there are three other numbers that control the power of the strong and weak nuclear forces and the electromagnetic force. Then we have the speed of light and the Planck constant, which fixes the size of atoms. After that, there are numbers that describe the mass of elementary particles, such as the proton, the electron, and so on. These constants play a fundamental role in how a universe evolves. They determine not only the mass and size of the galaxies, the stars, and the Earth, but also of living beings: the height of trees, the shape of a rose petal, the weight and size of ants, giraffes, and people. The reality we know would be quite different if the constants changed. As their name suggests, these constants do not vary in time or in space. This has been checked by careful observation of far-off galaxies. As for the initial conditions of the universe, they concern, among other things, the amount of matter it contains and its initial expansion rate.

If these constants and intitial conditions were just slightly different, then we wouldn't be here talking about them. The universe, right from the start, seems to have carried the seeds that allowed for the emergence of consciousness, of an observer. In the words of the physicist Freeman Dyson, "The universe in some sense must have known that we were coming."[3]

M: The fact that our existence seems to be coded into each part of the universe just shows how compatible we are with the physical world. But it certainly doesn't mean that you can say this is true because there was some intention that we should be here.

T: Yes, but so far we haven't come up with a theory that explains why these constants were fixed at a particular value and not a different one. We've been handed these numbers on a plate and have to accept them.

M: Is there really no explanation for this?

T: Leaving aside pure chance, which we shall come back to later, there is the superstring theory, which states that the elementary particles are created by the vibrations of infinitely tiny bits of strings. According to the theory, the mass and charge of a particle are determined by how the strings vibrate. But this doesn't really solve the problem, because the theory doesn't explain why the strings vibrate one way and not another, or why they create particles with exactly the properties needed.[4]

M: Could these physical constants be different in a different universe?

T: According to our present state of knowledge, there's no reason why they shouldn't vary from one universe to another. By constructing a large number of "model universes" on their computers, astrophysicists have discovered that if the physical constants and the initial conditions were just slightly different, then there'd be no life in the universe.

M: How much would they have to change by?

T: The exact figure depends on which constant and which initial condition we're talking about. But in each case, just a tiny change would make the universe barren. For instance, let's take the initial density of the matter in the universe. Matter has a gravitational pull that counteracts the force of expansion from the Big Bang and slows down the universe's rate of expansion. If the initial density had been too high, then the universe would have collapsed into itself after some relatively short time—a million years, a century, or even just a year, depending on the exact density.

Such a time span would have been too short for the nuclear alchemy of the stars to produce heavy elements like carbon, which are essential to life. On the other hand, if the initial density of matter had been too low, then there would not have been enough gravity for stars to form. And no stars, no heavy elements, and so no life! Everything hangs on an extremely delicate balance.

To give you an idea of just what small changes I mean, consider the density of the universe at the start (or at Planck time). It had to be fixed with an accuracy of around 10^{-60}. That is to say, if one figure after sixty zeros had been different, then the universe would be barren. There would be no life, no consciousness, and no you and me to discuss it. This astonishing precision is analogous to the dexterity of an archer hitting a one-centimeter-square target placed 15 billion light-years away, at the other side of the observable universe!

In the face of such extraordinary fine-tuning, many cosmologists have argued that the universe was so finely tuned in order to allow it to produce life, consciousness, and finally an intelligent observer capable of appreciating its beauty and harmony. This idea is called the "anthropic principle," from the Greek *anthropos,* meaning "person." According to this view, humanity has gained pride of place in the world once more—not at the center of the universe, but by being the very reason the universe was designed as it is.

This is the "strong" version of the anthropic principle. There is also a "weak" version that doesn't presuppose any intention in the design of nature. It almost comes down to a tautology: "The properties of the universe must be compatible with the existence of humankind."[5] The term "anthropic" is really inappropriate, because it implies that humanity in particular was the goal toward which the universe has evolved. But the arguments I've just presented can be applied to any form of intelligence in the universe.

Whichever of these versions of the principle one prefers, modern cosmology has, in this anthropic way of thinking, rediscovered the notion of a close link between humankind and the universe. Paul Claudel's message of hope answered Pascal's cry of despair and expressed this new delight with the world: "The silence of endless space no longer terrifies me. I walk through it with familiar ease. We do not live in some isolated part of a wild and hostile desert. Everything in the world is fraternal and familiar."

M: As far as Buddhism is concerned, the idea that there is some principle of organization that is supposed to have tuned the universe perfectly so that the conscious mind could evolve is fundamentally misguided. The apparently amazing fine-tuning is explained simply by the fact that the physical constants and consciousness have always coexisted in a universe that has no beginning and no end. I don't mean that the universe is static. Rather I mean that what seems to be the start of the universe, the Big Bang for instance, is just one episode in an unbroken process. The conditions of our present universe harmonize with those of the previous and subsequent ones, because the process of causality is unbroken and entails a compatibility between the nature of the cause and that of the effect.

The universe has not been adjusted by a great watchmaker *so that* consciousness can exist. The universe and consciousness have always coexisted and so cannot exclude each other. To coexist, phenomena must be mutually suitable. The problem with the anthropic principle, or any other teleological theory, is that it puts the constants before consciousness and thus claims that the constants exist only so that they can create consciousness. The anthropic principle comes down to picking up two halves of a walnut and saying, "It's incredible, it looks like these two pieces have been designed to fit perfectly together."

T: I can certainly see that Buddhism has no need of an anthropic principle to explain life and the conscious mind. But let's just suppose that the Buddhist view is invalid, and that it is, after all, necessary to explain the fine-tuning of the physical constants and the initial conditions of the universe. We might then wonder if this tuning was the consequence of chance or necessity, to quote Monod's book. If we do not want to accept the anthropic principle—the argument from necessity—then we would have to turn to the other main argument cosmologists use to explain the fine-tuning: the chance hypothesis.

This hypothesis postulates an infinite number of other universes, and that each of these universes is constructed according to one out of all of the possible combinations of physical constants and initial conditions. Thus for each combination of constants and initial conditions, a universe has once existed. But ours was the only universe born with just the right combination to have evolved to create life. All the others were losers and only ours is the winner. If you play the lottery an infinite number of times, then you inevitably end up winning the jackpot.

M: When you say "other universes," do you mean parallel ones?

T: That's one possibility. The idea of parallel universes is certainly one of the strangest notions in physics. The idea depends upon one of the fundamental, and quite strange, findings of quantum mechanics. According to quantum theory, it is impossible to state precisely where subatomic particles are located. We can only calculate the probability of finding an electron here or there; we cannot determine precisely its trajectory in the way that we can determine the orbit of a planet around the Sun. This lack of precise location is what's called "quantum fuzziness," and it leads to a discomfortingly strange, probabilistic description of the world. According to such a description, any given electron might be here, there, or anywhere all at once. In order to find a way around this and to revert to the more comforting realist view, where any given electron has only one position, the American physicist Hugh Everett came up with a radical idea: he proposed that the universe divides into as many nearly identical copies of itself as there are possibilities for the position of the electron. Instead of the electron being here, there, and everywhere in a single probabilistic universe, there would be one universe in which the electron would be here, another universe in which it would be there, yet a third universe in which it would be at another position, and so on. Through this continual process of division, a vast range of universes would be born. Some of those universes would evolve to be very similar to ours—maybe even differing by only the position of one electron in one atom. Others would differ more. There would be one in which you had gone for a stroll instead of sitting here talking to me. Other universes would exist in which Tibet hadn't been invaded by China and in which men hadn't walked on the Moon. Others still would be more radically different. Their physical constants, initial conditions, and physical laws wouldn't be the same. Each time the universe divided, all living beings would also be divided, you and I included. The parallel universes created would be isolated from one another. No communication between them would be possible.

M: When it comes to conscious beings, this would mean that our consciousness would be splitting in two at every single instant. The notion of an individual stream of consciousness would become meaningless.

T: That's an interesting argument against parallel universes! Personally, I find the idea that the universe divides into multiple universes in this way decidedly odd. It's not at all clear to me that our bodies and minds could divide in such a way without our noticing.

M: Buddhism considers that there is a plurality of worlds, or of states of being, but it doesn't agree that there could be totally disconnected parallel universes, living separate lives. Such an idea would run against the idea of the global interdependence of all phenomena. What can possibly be "outside" the infinite number of phenomena that have been linked together for all time and shall remain linked forever? If phenomena exist only in terms of interdependence, then we cannot separate them without causing them to disappear. Separately existing universes can't cohabit in such a global infinity.

Another problem with this hypothesis is that if you play the scenario out, you see that an infinity of universes is created, and with this infinity, all possible universes must exist, including ones in which totally contradictory conditions are true, or in which opposite events happen. But in a global interdependence, mutually exclusive phenomena cannot coexist. Additionally, if, at any moment, there were as many universes being created as there are particles in the universe (since these particles constantly change states, and so each new state would lead to at least two universes), such a process would destroy all idea of causality. For the same causes and the same initial conditions could *simultaneously* lead to opposite results.

T: That's exactly what this interpretation of quantum mechanics argues. The existence of parallel universes is dependent on what we call "true chance," that is to say the absence of causality.[6] In a world of parallel universes, the idea of moral responsibility becomes meaningless. This scenario argues that all possible outcomes of events must happen, each in its own universe. Therefore a criminal could plausibly argue that he should not be held accountable for his crime because this is the universe that just happens to be the one in which, by chance, he perpetrated it. And in addition, he could argue, there are many other versions of himself in other universes who did not commit the crime.

M: But if phenomena can exist without any cause, anything can result from anything. However, if phenomena are governed by global interdependence, then the world does not allow for such pure chance, and freedom itself can only be expressed through the laws of cause and effect.

T: I agree that the idea of parallel universes is implausible. But there is another way of thinking about multiple universes. We've already discussed the notion of cyclical universes, which would not be parallel in time, but successive. Whenever the universe was reborn from its ashes, it would start all over again with a new combination of physical constants and initial conditions. Almost all of these cycles would produce barren universes, incompatible with the appearance of life and consciousness. Then, from time to time, a universe like ours would crop up with a winning combination. The big problem with this idea is that, for the moment, astronomical observations seem to show that the universe doesn't contain enough matter for gravity to reverse the outward motion of the galaxies and pull them into a Big Crunch, which is necessary in order to kick off a new cycle of growth. According to our present state of knowledge, this expansion will be eternal.

Physicists are an imaginative lot, however, and they have come up with a scenario whereby a new Big Bang would not require a preceding Big Crunch. The American physicist Lee Smolin has proposed the idea that a new universe could spring out of the midst of a black hole[7] in a fantastic explosion just like our Big Bang, thus creating a new arena of time and space. The conditions at the time of the Big Bang and at the center of a black hole are similar because they are both characterized by extreme densities. The new universe created in this way would be totally disconnected from ours, because no information could pass through the center of a black hole and so reach our universe. So far, this scenario has no experimental backing, and is more like science fiction than science.

M: This new universe might be disconnected in terms of *transmission of information from one to the other,* but not in terms of causality, given that the black hole came from our universe. There would thus be a certain continuity between the two universes.

T: There is one other interesting version of the multiple universes idea. A Russian physicist, Andrei Linde, has proposed a theory whereby each of the infinite number of fluctuations of the primordial quantum foam created a universe. Our universe would then be just a tiny bubble in a superuniverse made up of an infinite number of other bubbles. None of those other universes would have intelligent life, because their physical constants and initial conditions wouldn't be suitable.

Intriguing as all of these notions are, I personally find the idea of multiple universes, parallel or otherwise, hard to swallow. The fact that all of these universes would be unobservable, and thus unverifiable, contradicts my view of science. Science rapidly becomes bogged down in metaphysics if it is not subjected to the test of experimental proof.

M: All the same, metaphysical preferences are constantly being used in scientific work! When several mutually exclusive hypotheses turn out to be equally good explanations for experimental evidence, then physicists are often influenced by metaphysical considerations when deciding which one to adopt. Declaring, for instance, that anything that is unobservable by physical means is nonexistent is a metaphysical point of view. Such inexistence lacks any proof.

In the end, it's impossible to eliminate metaphysics when dealing with our origins. No matter what science may discover about the Big Bang, the pre–Big Bang, or any other type of beginning for the universe, only metaphysics can ask, "Was there ever a beginning?" Or else, "Why should there be a beginning?" This is also the opinion of some scientists. François Jacob wrote: "One field must be totally excluded from scientific investigation. This is the origin of the world, the meaning of the human condition and human life's 'destiny.' Not that these are futile questions. Each of us asks them at some time or other. But these matters, which Karl Popper called *ultimate,* concern religion, metaphysics, or even poetry."[8]

T: Okay, I agree that my position is a metaphysical one. And yet it's always been observation that has ultimately decided the fate of scientific theories. For example, Einstein's theory of gravity has replaced Newton's because relativity can explain phenomena that escaped Newton. No matter how beautiful and harmonious the theory of relativity might be, it would not have been accepted without experimental demonstration. This position is, in a way, my version of Pascal's wager. It's also partly inspired by a principle of economy called Occam's razor, after the fourteenth-century theologian and philosopher William of Occam. The idea is to cut out all the hypotheses that are unnecessary when it comes to explaining a situation, and go for a simple answer, since it's more likely to be right than a complicated one. So why create an infinite number of barren universes just to have one that is conscious of its own existence?

M: Are you implying that the Creator did not have to produce a whole batch of failures for the purpose of coming up with the right universe, which contains us? This once again presupposes that there was a purpose in the creation of life and consciousness. But who is this Creator? Where did he come from? From a creator of creators? If not, then he must be his own cause. An effect, or a result, cannot be its own cause.

T: Another factor that sets me against the chance hypothesis is that I can't imagine that the world's profound beauty and harmony could have been produced at random. Our universe is beautiful—glowing red sunsets, the delicate outline of a rose petal, the wonderful pictures of stellar nurseries, or the elegance of a galaxy's spiral arms touch our very souls. The universe is harmonious because the laws that govern it don't vary either in time or in space.

M: The beauty argument also fails to stand up. Beauty is purely relative. A rose petal is a thing of beauty to a poet, food to an insect, and absolutely nothing to a whale. Spiral galaxies weren't beautiful to anyone until a small minority of people had the chance to examine them in the twentieth century. If the universe had been adjusted so that an observer capable of appreciating its beauty and harmony would appear, then it would be enough just to have a scattering of intelligent observers here and there in the universe. What would be the point of having beautiful uninhabited (or uninhabitable) planets, as well as galaxies that no one will ever see?

T: I grant you that at first sight such an argument seems itself to go against Occam's razor. But, in principle, all of those planets and galaxies could one day be observed by us or by an extraterrestrial intelligence.

I also have one more argument for my wager against chance. There is a profound unity in the universe. As physics progresses, it has unified phenomena that we used to think were totally distinct. In the seventeenth century, Newton unified Heaven and Earth. The same gravitational force makes an apple fall in an orchard or the Moon orbit around the Earth. In the nineteenth century, Maxwell showed that electricity and magnetism were just two different aspects of the same phenomenon. He then realized that electromagnetic waves were light waves, thus unifying optics with electromagnetism. At the beginning of the twentieth century, Einstein unified time and space, and at the dawn of the twenty-first century, physicists are doing their utmost to unify the four fundamental natural forces into one superforce. The physical laws that describe the universe tend toward unity. I therefore have a hard time with the idea that this profound unity is the result of pure chance.

M: The fact that the universe, and our own existence, might have evolved simply according to the laws of causality, without the intention of a Creator driving the process, would not exclude the possibility that the universe displays harmony, and also wouldn't imply that, as Jacques Monod wrote, "man is lost amidst the unfeeling immensity of the universe, where he arrived by pure chance," or that the universe is meaningless. We can find a great deal of meaning in the way our world is. The interdependence of the world and our consciousness allows us to use that consciousness to attain knowledge, which gives a very definite meaning to the whole setup.

T: I should clarify that I don't think the ultimate purpose or meaning of the universe is human life. Life surely hasn't finished its ascension toward complexity. The universe will continue to evolve, and humankind with it. My point is that I have problems with the idea that this cosmic evolution leading to human life came about purely through a series of wonderful coincidences, of lucky rolls of the dice that could easily never have happened. I should be clear that I have no problem with the existence of the kind of chance that Jacques Monod lamented. When a biochemist like him talked about chance, he meant the chance encounters between quarks in order to form the nuclei of atoms, between atoms at the center of the stars in order to feed their fires, between the atoms created by stellar combustion in order to form interstellar molecules and planets, and between organic molecules in the primordial oceans in order to form the interwoven helices of DNA. I have no doubt that these processes are, in a fundamental way, ruled by the chance revealed by quantum theory. In this sense, I can agree that contingency plays an important role in the evolution of the universe. I certainly do not subscribe to Laplace's view of a completely deterministic universe.

It would be absurd to say that the first moments of the universe totally determined the fact that we are now here talking. Rather, I think that as soon as the laws of physics were fixed, they formed a cloth that was ideally suited for nature to embroider. The manner in which that embroidery proceeds is one driven largely by chance. Quantum fluctuations in the world of atoms, chaotic phenomena in the macroscopic world such as those that are responsible for the weather, as well as accidental, contingent events—such as the asteroid that hit the earth 65 million years ago and killed the dinosaurs, thus allowing the emergence of mammals and subsequently human beings—have all freed nature from its deterministic straitjacket. Nature has evolved to be the inventor of complexity. Just as a jazz musician embroiders around a theme, thus improvising new melodic phrases according to his inspiration and the public's reactions, nature plays spontaneously with the physical laws that were fixed at the start of the universe. It uses them to create novelty.[9] But

I cannot accept that the choice of physical constants and the initial conditions were matters purely of chance. As soon as the constants and the initial conditions were fixed, matter already contained the seeds of consciousness, and cosmic gestation started leading inexorably toward us.

In the end, if we reject chance and the idea of multiple universes, and if we postulate the existence of just one universe, ours, then it seems to me that, like Pascal, we must wager on the existence of a creative principle.

M: Very well. Let's now examine this principle. First, in your view, does it imply a will to create?

T: My view is that the constants and initial conditions were intentionally fixed so that they led to a universe that was conscious of itself. It's up to you if you want to call the cause of this selection God or not. I do not personally believe in a personified God, but rather in a pantheistic principle that is omnipresent in nature. This is somewhat akin to the views of Einstein and Spinoza. Einstein described it as follows: "The scientist is possessed by the sense of universal causation. . . . His religious feeling takes the form of a rapturous amazement at the harmony of natural law, which reveals an intelligence of such superiority that, compared with it, all the systematic thinking and acting of human beings is an utterly insignificant reflection."[10] He added: "I believe in Spinoza's God who reveals himself in the harmony of all that exists, but not in a God who concerns himself with the fate and actions of human beings."[11]

Modern science has in fact provided a good deal of new evidence against several of the arguments that Western philosophers and theologians have used to prove the existence of a "classic" God. One of these arguments in proof of God is what I'll call the complexity argument, which states that only a Creator could have produced such a complex, structured universe. A watch has to be made by a watchmaker; it can't assemble itself. You can't write a book just by throwing an inkpot, a pen, and some sheets of paper onto a table. But contemporary science has shown that highly complex systems can in fact result from a perfectly natural evolution based on known physical and biological laws, and there is no need to call in a watchmaker God.

Then there is the "cosmological" argument, which was used by Plato, Aristotle, Saint Thomas Aquinas, and Kant. This argument states that everything has a cause, but that there can't be an infinite chain of causes. The chain must necessarily begin with a prime cause, and they argued that that prime cause was God. Quantum mechanics has raised doubts about the premise that "everything has a cause." Thanks to his famous "uncertainty principle," Werner Heisenberg showed in 1927 that uncertainty is an inherent part of the subatomic world and that particles can materialize in unpredictable ways, without the slightest need for any cause.

Just like any particle, the universe can, in principle, spring up from a vacuum, without any prime cause, thanks to a quantum fluctuation.

M: Why shouldn't a chain of causes be infinite in time and complexity? What law of nature does that contradict? How many causes must we have before saying, "That's enough, I can't keep going back in time ad infinitum, so let's adopt a causeless creator"?

T: Yes, of course, the cosmological argument does depend on a linear concept of time. Other philosophers have suggested that time is circular, thus eliminating the need for a prime cause. So, instead of there being a linear progression in which *a* happens, which causes *b,* which leads to *c,* and so on, we would have a cyclical progression in which *a* causes *b,* which leads to *c,* which in turn is responsible for *a.* The snake bites its own tail and the cycle is closed. There's no more need for a prime cause!

M: Another odd idea! Like the prime cause, the idea of a closed circle is just an escape route for people who can't bear the idea of infinity. When you say that a particle materializes without any cause, you're limiting causality to a simple logic of *a* producing *b.* But if, through interdependence, *b* results from the existence of the entire universe—from an infinite number of causes and fluctuating relationships—we can't say that it has occurred spontaneously just because we are incapable of identifying a particular cause.

T: Another scientific argument against the existence of God is that the very idea of cause and effect loses its meaning when applied to the universe. This notion presupposes the existence of time, so that cause reliably precedes effect. But according to the Big Bang, time and space appeared simultaneously with the universe. If time didn't exist before this, then what does "and God created the universe" mean? The act of creating the universe is meaningful only in time. Is God in time, or outside of it? Time isn't absolute, as Einstein said. It's elastic and is stretched (or contracted) by accelerating motions or fields of intense gravity, such as those around black holes. A God contained in time would no longer be all-powerful, because he would be subject to the laws of time. A God outside time would be omnipotent, but unable to help us, since our actions happen in time. If God transcended time, then he would already know the future. If he knew everything in advance, why would he bother to become involved in the struggle of humankind against evil?

M: God must be either immutable, and thus unable to create, or else inside time and thus not immutable. This is one of the contradictions that the notion of a prime cause leads to. What are the justifications behind this argument?

First, if there *is* a prime cause, it should be immutable. Why? Because, by definition, it has no other cause than itself, so it has no reason to become different. Change would imply the intervention of another cause that wasn't a part of the prime cause.

Second, how could an immutable entity create something? If there is an act of creation, is the creator involved or not? If he is not, why call him "creator"? If he is involved, then because creation inevitably occurs in stages, the something or someone involved in these stages is not immutable. One could agree with Saint Augustine that God created time and the universe. But even so, creation remains a process, and any process, whether temporal or not, is incompatible with immutability. This point did not escape Saint Augustine himself, who said that the notion of beginning involves an act of faith. Buddhism contends, by contrast, that such an act of faith is unnecessary provided one doesn't cling to the position that there must have been a beginning.

T: So there are many compelling arguments against the existence of a God. But the principle of organization that I believe in is quite different from a "classic" God. What I'm talking about is a principle that finely tuned the universe at the beginning, and not a personified God.

M: But if you talk about a principle of organization, you can't be so vague about it. We must be able to discuss it.

T: These questions have endlessly bothered theologians and philosophers of all persuasions, and I certainly don't claim to have the answers. I am engaged here in a metaphysical wager. Science has no special authority about this subject.

M: But we can't postulate the existence of something and at the same time have nothing to say about it. If the principle of organization has no characteristics, then it doesn't matter whether it exists or not. We can eliminate it with Occam's razor. Was this principle born of itself? Was it caused by something? Is it permanent? Omnipotent?

T: Omniscience, omnipotence, and so on are the qualities of a God. And it's true that we tend to identify this principle of creation with a Creator God. Physicists don't talk about God, however; rather, they refer to physical laws. The properties of those laws are strangely reminiscent of how God is generally described, and this leads many people to see a relationship between them. The physical laws are, for example, universal and are applied everywhere in time and space, from the tiniest atom to the largest galaxy. They're absolute, since they don't depend on the person that discovers them. The fact that discoveries are often made by individuals does not mean that the individuals have created the truths they find. The truths are there to be found. The physical laws are also timeless, because even though they describe a world governed by time and constantly changing phenomena, they themselves never change. We live in a temporal universe ruled by atemporal laws. The laws are also omnipotent, because they apply to everything everywhere. Finally, they're omniscient, because they act on material objects without having to be "informed" of the particular states of these objects. They "know" in advance and "legislate" just the right

instructions to "command" the behavior appropriate to those states. And so the characteristics of physical laws are the same as God's.

M: I still don't see why these laws should point to the existence of a creative or organizing principle. They quite simply reflect the interdependent nature of phenomena.

T: I agree that the idea of interdependence could explain the fine-tuning of the physical laws and initial conditions that allowed life and consciousness to come about. But I don't see how interdependence answers Leibniz's existential question, "Why is there something rather than nothing?"

M: But postulating the existence of a principle of organization doesn't answer this question either. Why was there a principle of organization rather than nothing? And what do we mean by a principle of organization? Is this principle some kind of an entity, some kind of consciousness that conceived a functional model of the universe? Did this principle *decide* that it wanted to create? Does it have intention?

T: I believe the answer is yes, the principle of organization wanted to create a conscious, intelligent observer. I believe this is why our universe was set up to evolve in the manner it has.

M: But if the principle did decide to create, then we must recognize that it cannot be all-powerful, because it was influenced by the desire to create. If you then argue the other way and say that the principle did not actually decide to create, then you must concede that the principle is not all-powerful because it created without deciding to create. Therefore it wasn't free to create or not to create.

This principle also cannot be timeless, or immutable, because, as we have seen, in the process of creating it will necessarily have itself changed. It went from having the desire to create to having done so, and therefore it would no longer have the desire in the same way. Creation implies change—there is a creative act. After creating, the principle would no longer be the same—before it wasn't a creator and after it was. The principle has thereby lost its immutability.

Another troubling question about this principle is whether it was born without a cause, or whether it was perhaps its own cause.

T: I think that it was its own cause. But, once again, we are stepping out of the domain of science here, and this is just a matter of my beliefs.

M: If the principle was its own cause, then, once again, Buddhism would argue that it must be immutable. An entity that exists because of itself has no reason to change. Only things produced by something else can change. So we run into another contradiction, because if the principle is immutable, then it can't create. Buddhism argues that something permanent can't produce something ephemeral. What's more, as said above, if it creates, then it is no longer immutable, since creation implies change. Any creator must in turn be modified by its creation, because each action implies an interaction. No cause can be one-way. Causality is necessarily reciprocal. Any creator is acted upon by its creation.

Another very important point is that if something has no cause other than itself, the concept of interdependence says that it is then prohibited from interacting with other things. Buddhism does not believe that anything can be the cause of itself.

T: So each event or each entity must have a cause. Doesn't that lead to an infinite regression?

M: Such a form of infinity definitely goes against traditional Western metaphysical beliefs. Both the religious and scientific viewpoints have insisted that there must be an ultimate "beginning." As the philosopher Bertrand Russell wrote, "There is no reason to suppose that the world had a beginning at all. The idea that things must have a beginning is due to the poverty of our imagination."[12] This desire to find a beginning comes from the idea that everything has the real, solid existence that our minds generally perceive.

T: And yet, as I explained earlier, quantum mechanics does provide a way around this need for an ultimate beginning. A "beginning" is no longer really necessary.

M: If a beginning is no longer necessary—and there I'm in complete agreement—then neither is a principle of organization. How could it have organized phenomena that have no beginning? All it could do would be to modify how they evolve.

Another argument against this principle of organization is that if it organized the entire world of phenomena, then it must contain all of the causes of that world. Otherwise, something would exist that was outside its creation.

T: That's logical.

M: But the law of causality says that if an event doesn't take place, then it's because certain causes or conditions were missing. If a seed doesn't germinate, then it must be defective, or else lack water, heat, and so on. An effect can't occur until all the necessary causes and conditions have come together. On the other hand, when they are all present, then the effect must necessarily occur.[13] If it doesn't happen, then something must still be missing. So, if this principle of creation contained all of the universe's causes and conditions, it must constantly create the entire universe. It would be something like a permanent Big Bang.

T: You mean that it couldn't stop?

M: That's right. If it did stop, then this would mean that it no longer contained all of the causes and conditions of creation. It would then need help, either from another principle upon which it was dependent because this second principle had some of the causes, or else from a regulating principle that checked its creation. Either way, it would lose its omnipotence. So there are just two possibilities: either it doesn't contain all the causes and conditions, and so cannot create; or it does have them all, and so creates continuously.

T: Which is also absurd.

M: Some people counter this objection by saying that the creator could create the world progressively. That, too, would imply that the universe had multiple causes and that the creator didn't possess all of the causes at the beginning of the process. So where did it subsequently get them from?

Others claim that the act of creation is atemporal, that the world is being created now, just as much as it has been and will be created. This is to see creation as a series of events that are simultaneous from the Creator's viewpoint. But this position doesn't escape the problem of a principle that must contain all the causes of the universe and so continually create the entire universe. What's more, if all the universe's past, present, and future events were simultaneous so far as the Creator was concerned, then this would lead to absolute determinism. It would mean that any attempt at personal transformation in order to dissipate ignorance and reach enlightenment would be in vain. On the other hand, absolute determinism can be avoided by the play of interdependent relationships, without a beginning or end, because the causes and conditions of these relationships are unlimited and so can't be reduced to a single prime cause.

One final argument against the whole notion of a beginning or creation is that if the Creator possessed not only all the causes but also all the effects, then we could hardly talk about creation anymore, since everything was already there.

Some Hindu philosophers have said that the Creator becomes the matching cause for each step in the creation while still remaining immutable, as a dancer can perform different choreographies in different costumes and still remain the same person. But that contradicts the idea that the Creator is the sole cause of the universe: a single cause can't lead to diverse effects. What is more, a unity with changing aspects can neither be a true unity nor immutable. If the effect is intermittent, then the cause can't be eternal.

To sum up the Buddhist alternative way of thinking, which requires no principle of organization, in the Buddhist world of appearances, each instant is a perpetual end and beginning because of the basic impermanence of the phenomena produced by the laws of cause and effect. In terms of absolute truth, all past, present, and future events are identical in that they have no intrinsic existence. Thus they have no real end or beginning. If nothing is really "produced," there is no need to look for an end. And so it isn't necessary to search for a principle of organization that is supposed to have made everything and have been made only by itself.

T: So, flying in the face of monotheistic religions, Buddhism categorically denies the existence of God the Creator. Doesn't that sit uneasily with the image of tolerance that Buddhism usually likes to project? How do you reconcile this position with respect for other beliefs?

M: No matter how tolerant you are, you don't have to accept other people's metaphysical viewpoints. But you must respect the means of personal transformation that suit best other people's natures and predispositions. Also, there are many ways of thinking about God. To quote the Dalai Lama, there seems to be "a way of looking at God not so much in terms of a personal deity but rather as a ground of being. Yet qualities such as compassion can also be attributed to that divine ground of being. Now if we are to understand God in such terms—as an ultimate ground of being—then it becomes possible to draw parallels with certain elements in Buddhist thought and practice."[14] But neither should we make an amalgam of all the forms of religion, spirituality, science, humanism, or agnosticism! That isn't the aim of tolerance. The Dalai Lama went on, "I do not personally advocate seeking a universal religion; I don't think it advisable to do so. And if we proceed too far in drawing these parallels and ignoring the differences, we might end up doing exactly that!"[15]

Metaphysical positions must be clearly expressed. There is no reason to be ambiguous about them. If they're wrong, then let's prove it. Buddhism is quite prepared to admit its own mistakes, if they can be proven to be so. Intolerance consists in being so sure of the truth that you want to impose it on everyone else by persuasion, or even by force. We must have an open mind to realize that what suits us doesn't necessarily suit others. "One cannot eat a particular food and then say, 'Because it is nutritious for me, everyone must eat it'; each person must eat foods that are suitable for the best physical health according to his or her own physical constitution."[16]

As regards spiritual practice, belief in God can give some people a feeling of closeness with their Creator and encourage them to become more loving and altruistic in order to express their thankfulness and participate in the love of God for living creatures. For others, a deep understanding of interdependence and the laws of cause and effect, linked with a desire to reach enlightenment in order to help all sentient beings, are greater sources of inspiration in the development of love and compassion. To conclude: "When embarking upon a spiritual path, it is important that you engage in a practice that is most suited to your mental development, your dispositions, and your spiritual inclinations. . . . Through this, one can bring about inner transformation, the inner tranquility that will make that individual spiritually mature and a warm-hearted, whole, and good and kind person. That is the consideration one must use in seeking spiritual nourishment."[17]

THE UNIVERSE IN A GRAIN OF SAND

4

THE INTERDEPENDENCE AND NONSEPARABILITY OF PHENOMENA

The concept of interdependence lies at the heart of the Buddhist vision of the nature of reality, and has immense implications in Buddhism regarding how we should live our lives. This concept of interdependence is strikingly similar to the concept of nonseparability in quantum physics. Both concepts lead us to ask a question that is both simple and fundamental: Can a "thing," or a "phenomenon," exist autonomously? If not, in what way and to what degree are the universe's phenomena interconnected? If things do not exist per se, what conclusions must be drawn about life?

THUAN: As we learned in the last chapter, Buddhism rejects the idea of a principle of creation, as well as the radical notion of parallel universes—though it may accommodate the idea of multiple universes. To Buddhism, the extraordinary fine-tuning of the physical constants and the initial conditions that allowed the universe to create life and consciousness are explained by "the interdependence of phenomena." I think it's time to explain more about this idea.

MATTHIEU: To do so, we should first return to the concept of "relative truth." In Buddhism, the perception we have of distinct phenomena resulting from isolated causes and conditions is called "relative truth" or "delusion." Our daily experience makes us think that things have a real, objective independence, as though they existed all on their own and had intrinsic identities. But this way of seeing phenomena is just a mental construct. Even though this view of reality seems to be commonsense, it doesn't stand up to analysis.

Buddhism instead adopts the notion that all things exist only in relationship to others, the idea of mutual causality. An event can happen only because it's dependent on other factors.[1] Buddhism sees the world as a vast flow of events that are linked together and participate in one another. The way we perceive this flow crystallizes certain aspects of the nonseparable universe, thus creating an illusion that there are autonomous entities completely separate from us.

In one of his sermons, the Buddha described reality as a display of pearls—each pearl reflects all of the others, as well as the palace whose façade they decorate, and the entirety of the universe. This comes down to saying that all of reality is present in each of its parts. This image is a good illustration of interdependence, which states that no entity independent of the whole can exist anywhere in the universe.

T: This "flow of events" idea is similar to the view of reality that derives from modern cosmology. From the smallest atom up to the universe in its entirety, including the galaxies, stars, and humankind, everything is moving and evolving. Nothing is immutable.

M: Not only do things move, but we see them as "things" only because we are viewing them from a particular angle. We mustn't give the world properties that are merely appearances. Phenomena are simply events that happen in certain circumstances. Buddhism doesn't deny conventional truth—the sort that ordinary people perceive or the scientist detects. It doesn't contest the laws of cause and effect, or the laws of physics and mathematics. It quite simply affirms that, if we dig deep enough, there is a difference between the way we see the world and the way it really is, and the way it really is, we've discovered, is devoid of intrinsic existence.

T: So what has that true nature got to do with interdependence?

M: The word "interdependence" is a translation of the Sanskrit *pratitya samutpada,* which means "to be by co-emergence" and is usually translated as "dependent origination." The saying can be interpreted in two complementary ways. The first is "*this* arises because *that* is," which comes down to saying that things do exist in some way, but nothing exists on its own. The second is "*this,* having been produced, produces *that,*" which means that nothing can be its own cause. Or we could say that everything is in some way interdependent with the world. We do not deny that phenomena really do occur, but we argue that they are "dependent," that they don't exist in an autonomous way. Any given thing in our world can appear only because it's connected, conditioned and in turn conditioning, co-present and co-operating in constant transformation. Their way of "being" is simply in relation to one another, never in and of themselves. We tend to cling to the notion that "things" must precede relationships. This is not the case here. The characteristics of phenomena are defined only through relationships.

Interdependence explains what Buddhism sees as the impermanence and emptiness of phenomena, and this emptiness is what we mean by the lack of "reality." The seventh Dalai Lama summarized this idea in a verse:

> *Understanding interdependence, we understand emptiness*
> *Understanding emptiness, we understand interdependence.*
> *This is the view that lies in the middle,*
> *And which is beyond the terrifying cliffs of eternalism and nihilism.*[2]

Another way of defining the idea of interdependence is summarized by the term *tantra,* which stands for a notion of continuity and "the fact that everything is part of the whole, so that nothing can happen separately."[3]

Ironically, though we might think that the idea of interdependence undermines the notion of reality, in the Buddhist way of thinking, it is interdependence that actually allows for reality to appear. Let's think about an entity that exists independently from all others. As an immutable and autonomous entity, it couldn't act on anything, or be acted on itself. For phenomena to happen, interdependence is required.

This argument refutes the idea of distinct particles that are supposed to constitute matter. What's more, this interdependence naturally includes consciousness. The reality of any given object depends on a subject that is aware of that object. This was what the physicist Erwin Schrödinger meant when he wrote: "Without being aware of it, and without being rigorously systematic about it, we exclude the subject of cognizance from the domain of nature that we endeavor to understand. We step with our own person back into the part of an onlooker who does not belong to the world, which by this very procedure becomes an objective world."[4]

Finally, the most subtle aspect of interdependence, or "dependent origination," concerns what we call a phenomenon's "designation base" and its "designation." A phenomenon's position, form, dimension, color, or any other of its apparent characteristics is merely one of its "designation bases." This designation is a mental construct that invests a phenomenon with a distinct reality. In our everyday experience, when we see an object, we aren't struck by its nominal existence, but rather by its *true* existence.[5] If we analyze this "object" more closely, however, we discover that it is produced by a large number of causes and conditions, and that we are incapable of pinpointing an autonomous identity. Since we have experienced it, we can't say that the phenomenon doesn't exist. But neither can we say that it corresponds to an intrinsic reality. So we conclude that the object exists (thus avoiding a nihilistic view), but that this existence is purely nominal, or conventional (thus also avoiding the opposite extreme of material realism, which is called "eternalism" in Buddhism). A phenomenon with no autonomous existence, but that is nevertheless not totally inexistent, can act and function according to causality and thus lead to positive or negative effects. This view of reality therefore allows us to anticipate the results of our actions and organize our relationship with the world. A Tibetan poem puts it this way:

To say a thing is empty does not mean
It cannot function—it means it lacks an absolute reality.

To say a thing arises "in dependence" does not mean
It has intrinsic being—it means it is illusion-like.
If thus one's understanding is correct and certain
Of what is meant by voidness and dependent origin,
No need is there to add that voidness and appearance
Occur together without contradiction in a single thing.

T: I find everything you've told me about interdependence striking. Science, too, has discovered that reality is nonseparable, or interdependent, both at the subatomic level and in the macrocosmic world. The conclusion that subatomic phenomena are interdependent was derived from a famous thought experiment conducted by Einstein and two of his Princeton colleagues, Boris Podolsky and Nathan Rosen, in 1935. It's called the EPR experiment, from the initials of their surnames.

To follow this experiment, you need to know that light (and matter, too) has a dual nature. The particles we call "photons" and "electrons," as well as all the other particles of matter, are Janus-faced. Sometimes they appear as particles, but they can also appear as waves. This is one of the strangest and most counterintuitive findings of quantum theory. Even stranger is the finding that what makes the difference about whether a particle is in the wave or particle state is the role of an observer—if we try to observe the particle in its wave state, it becomes a particle. But if it is unobserved, it remains in the wave state.

Take the case of a photon. If it appears as a wave, then quantum physics says that it spreads out in all directions through space, like the ripples made by a pebble thrown into a pond. The photon in this state has no fixed location or trajectory. We can then say that the photon is present everywhere at the same time. Quantum mechanics states that when a photon is in this wave state, we can never predict where the photon will be at any given moment; all we can do is evaluate the probability of its being in a particular position. The chances might be 75 percent or 90 percent, but never 100 percent. Since Einstein was a committed determinist, he couldn't accept that the quantum world was ruled in this way by probability or chance. He argued famously that "God does not play dice," and stubbornly set about trying to find the weak link in quantum mechanics and its probabilistic interpretation of reality. That's why he came up with the EPR experiment.

The experiment goes like this: First imagine that you have constructed a measuring apparatus with which you can observe the behavior of particles of light, called photons. Now imagine a particle that disintegrates spontaneously into two photons, *a* and *b*. The law of symmetry dictates that they will always travel in opposite directions. If *a* goes northward, then we will detect *b* to the south. So far, so good. But we're forgetting the strangeness of quantum mechanics. Before being captured by the detector, if quantum mechanics is correct, *a* appeared as a wave, not a particle. This wave wasn't localized, and there was a certain probability that *a* might be found in any given direction. It's only when it has been captured that *a* changes into a particle and "learns" that it's heading northward. But if *a* didn't "know" before being captured which direction it had taken, how could *b* have "guessed" what *a* was doing and ordered its behavior accordingly so that it could be captured at the same moment in the opposite direction? This is impossible, unless we admit that *a* can inform *b* instantaneously of the direction it has taken. But Einstein's cherished theory of relativity states that nothing can travel faster than light. The information about *a*'s location would need to travel faster than the speed of light in order to get to *b* in time, because, after all, *a* and *b* are both particles of light and are therefore traveling themselves at the speed of light. "God does not send telepathic signals," Einstein said, adding, "There can be no spooky action at a distance."

On the basis of these thought-experiment results, Einstein concluded that quantum mechanics didn't provide a complete description of reality. In his opinion, the idea that *a* could instantaneously inform *b* of its position was absurd: *a* must know which direction it was going to take, and tell *b* before they split up; *a* must then have an objective reality, independent of actual observation. Thus the probabilistic interpretation of quantum mechanics, which states that *a* could be going in any direction, must be wrong. Quantum uncertainty must hide a deeper, intrinsic determinism. Einstein thought that a particle's speed and position, which defined its trajectory, were *localized* on the particle without any observation being necessary. This is what was called "local realism." Quantum mechanics couldn't describe a particle's trajectory because it didn't take other "hidden variables" into account. And so it must be incomplete.

And yet Einstein was wrong. Eventually, physicists showed that exactly what Einstein thought couldn't happen in the EPR experiment did happen. Since its invention, quantum mechanics—and its probabilistic interpretation of reality—has never slipped up. It has always been confirmed by experiments and it still remains today the best theory that we have to describe the atomic and subatomic world.

M: When was the EPR effect confirmed experimentally?

T: EPR remained only a thought experiment for some time. No one knew how to carry it out physically. Then, in 1964, John Bell, an Irish physicist working at CERN, devised a mathematical theorem called "Bell's inequality," which would be capable of being verified experimentally if particles really did have hidden variables, as Einstein thought. This theorem at last allowed us to take the debate from the metaphysical plane to concrete experimentation. In 1982 the French physicist Alain Aspect, and his team at the University of Orsay, carried out a series of experiments on pairs of photons in order to test the EPR paradox. They found that Bell's inequality was violated without exception. Einstein had it wrong, and quantum mechanics was right. In Aspect's experiment, photons *a* and *b* were thirteen yards apart, yet *b* always "knew" instantaneously what *a* was doing, and reacted accordingly.

M: How do we know that this happens instantaneously, and that a light beam hasn't relayed the information from *a* to *b*?

T: Atomic clocks, connected to the detectors that capture *a* and *b*, allow us to gauge the moment of each photon's arrival extremely accurately. The difference between the two arrival times is less than a few tenths of a billionth of a second—it is probably zero, in fact, but existing atomic clocks don't allow us to measure periods of under 10^{-10} seconds. Now, in 10^{-10} seconds, light can travel only just over an inch—far less than the thirteen yards separating *a* from *b*. What is more, the result is the same if the distance between the two photons is increased. Even though light can definitely not have had the time to cross this distance and relay the necessary information, the behavior of *a* is always exactly correlated with that of *b*.[6]

The latest experiment was carried out in 1998 in Geneva by Nicolas Gisin and his colleagues. They began by producing a pair of photons, one of which was then sent through a fiber-optic cable toward the north of the city, and the other toward the south. The two pieces of measuring equipment were over six miles apart. Once they arrived at the end of the cables, the two photons had to choose at random between two possible routes—one short, the other long. It was observed that they always made the same decision. On average, they chose the long route half the time, and the short route half the time, but the choices were always identical. The Swiss physicists were sure that the two photons couldn't communicate by means of light, because the difference between their response times was under three-tenths of a billionth of a second, and in that time light could have crossed just three and half inches of the six miles separating the two photons. Classic physics states that because they can't communicate, the choices of the two photons must be totally independent. But that is not what happens. They are always perfectly correlated. How can we explain why *b* immediately "knows" what *a* is doing? But this is paradoxical only if, like Einstein, we think that reality is cut up and localized in each photon. The problem goes away if we admit that *a* and *b* are part of a nonseparable reality, no matter how far apart they are. In that case, *a* doesn't need to send a signal to *b* because these two light particles (or, rather, phenomena that the detector sees as light particles) stay constantly in touch through some mysterious interaction. Wherever it happens to be, particle *b* continues to share the reality of particle *a*.

M: Even if the two particles were at opposite ends of the universe?

T: Yes. Quantum mechanics thus eliminates all idea of locality. It provides a holistic idea of space. The notions of "here" and "there" become meaningless, because "here" is identical to "there." This is the definition of what physicists call "nonseparability."

M: This should have enormous repercussions on how physicists understand reality and our own ordinary perception of the world.

T: Indeed. Some physicists have had problems accepting the idea of a nonseparable reality and have tried to find a weak link in these experiments or in Bell's theorem. So far, they've all failed. Quantum mechanics has never been found to be wrong. So phenomena do seem "interdependent" at a subatomic level, to use the Buddhist term.

Another fascinating and famous experiment in physics shows that interdependence isn't limited to the world of particles, but applies also to the entire universe , or in other words that interdependence is true of the macrocosm as well as the microcosm. This is the experiment often referred to in short as Foucault's pendulum.

A French physicist, Léon Foucault, wanted to prove that the Earth rotates on its axis. In 1851 he carried out a famous experiment that is reproduced today in displays in many of the world's science museums. He hung a pendulum from the roof of the Panthéon in Paris. Once in motion, this pendulum behaved in a strange way. As time passed, it always gradually changed the direction in which it was swinging. If it was set swinging in a north-south direction, after a few hours it was swinging east-west. From calculations, we know that if the pendulum were placed at either one of the poles, then it would turn completely around in twenty-four hours. But because of the latitude of Paris, Foucault's pendulum performed only part of a complete rotation each day.

Why did the direction change? Foucault answered by saying that the movement was illusory. In fact, the pendulum always swung in the same direction, and it was the Earth that turned. Once he'd proved that the Earth rotated, he let the matter drop. But Foucault's answer was incomplete, because a movement can be described only in comparison with a fixed reference point; absolute movement doesn't exist. Long before, Galileo said that "movement is as nothing." He understood that it exists only relative to something else. The earth must "turn" in relation to something that doesn't turn. But where to find this "something"? In order to test the immobility of a given reference point, a star for instance, we simply set the pendulum swinging in the star's direction. If the star is motionless, then the pendulum will always swing toward it. If the star moves, then the star will slowly shift away from the pendulum's swing.

Let's try the experiment with known celestial bodies, both near and far. If we point the pendulum toward the Sun, after a few weeks there is a clear shift of the Sun away from the pendulum's swing. After a couple of years, the same happens with the nearest stars, situated a few light-years away. The Andromeda galaxy, which is 2 million light-years away, moves away more slowly, but does shift. The time spent in line with the pendulum's swing grows longer and the shift away tends toward zero the greater the distance is. Only the most distant galaxies, situated at the edge of the known universe, billions of light-years away, do not drift away from the initial plane of the pendulum's swing.

The conclusion we must draw is extraordinary: Foucault's pendulum doesn't base its behavior on its local environment, but rather on the most distant galaxies, or, more accurately, on the entire universe, given that practically all visible matter is to be found in distant galaxies and not in nearby stars. Thus, what happens here on our Earth is decided by all the vast cosmos. What occurs on our tiny planet depends on all of the universe's structures.

Why does Foucault's pendulum behave like this? We don't know. Ernst Mach, the Austrian philosopher and physicist who gave his name to the unit of supersonic speed, thought it could be explained by a sort of omnipresence of matter and of its influence. In his opinion, an object's mass—that is to say, the amount of its inertia, or resistance to movement—comes from the influence of the entire universe. This is what is called Mach's principle. When we have trouble pushing a car, its resistance to being moved has been created by the whole universe. Mach never explained this mysterious universal influence in detail, which is different from gravity, and no one has managed to do so since. Just as the EPR experiment forces us to accept that interactions exist in the microcosm that are different from those described by known physics, Foucault's pendulum does the same for the macrocosm. Such interactions are not based on force or an exchange of energy, and they connect the entire universe. Each part contains the whole, and each part depends on all the other parts.

M: In Buddhist terms, that's a good definition of interdependence. It's not a question of proximity in time or space, or of the speed of communication and physical forces whose influence wanes over great distances. Phenomena are interdependent because they *coexist* in a global reality, which functions according to mutual causality. Phenomena are naturally simultaneous because one implies the presence of the other. We are back with "*this* can only be if *that* also exists; *this* can change only if *that* also changes." Thus we arrive at an idea that everything must be connected to everything else. Relationships determine our reality, the conditions of our existence, particles and galaxies.

T: Such a vision of interdependence certainly agrees with the results of the experiments I've just mentioned. The EPR experiment, Foucault's pendulum, and Mach's inertia can't be explained by the four fundamental physical forces. This is extremely disturbing for physicists.

M: I think that we have a good example here of the difference between the scientific approach and Buddhism. For most scientists, even if the global nature of phenomena has been demonstrated in rather a disturbing way, this is merely another piece of information, and no matter how intellectually stimulating it may be, it has little effect on their daily lives. For Buddhists, on the other hand, the repercussions of the interdependence of phenomena are far greater.

The notion of interdependence makes us question our basic perception of the world and then use this new perception again and again to lessen our attachments, our fears, and our aversions. An understanding of interdependence should demolish the wall of illusions that our minds have built up between "me" and "the other." It makes a nonsense of pride, jealousy, greed, and malice. If not only all inert things but also all living beings are connected, then we should feel deeply concerned about the happiness and suffering of others. The attempt to build our happiness on others' misery is not just amoral, it's also unrealistic. Feelings of universal love (which Buddhism defines as the desire for all beings to experience happiness and to know its cause) and of compassion (the desire for all beings to be freed of suffering and its causes) are the direct consequences of interdependence. Thus knowledge of interdependence leads to a process of inner transformation, which continues throughout the journey of spiritual enlightenment. For, if we don't put our knowledge into practice, we are like a deaf musician, or a swimmer who dies of thirst for fear of drowning if he drinks.

T: So the interdependence of phenomena equals universal responsibility. What a marvelous equation! It reminds me of what Einstein said: "A human being is part of a whole, called by us the 'Universe,' a part limited in time and space. He experiences himself, his thoughts and feelings, as something separated from the rest—a kind of optical delusion of his consciousness. This delusion is a kind of prison for us, restricting us to our personal desires and to affection for a few persons nearest us. Our task must be to free ourselves from this prison by widening our circles of compassion to embrace all living creatures and the whole of nature in its beauty."

In fact, the language of physics is currently incapable of expressing the global, holistic nature of reality. Some people even talk of another truth, a "veiled reality" in the words of the French physicist Bernard d'Espagnat.[7]

M: This is an interesting idea, as long as we don't see this "veiled reality" as the ultimate solid reality hidden behind appearances. Doing so would just reify the world of phenomena once more. An important point we must keep in mind about interdependence is that it is not just a simple interaction between phenomena. Instead, it is the precondition for their appearance.

T: Heisenberg expressed a similar idea when he wrote, "The world thus appears as a complicated tissue of events, in which connections of different kinds alternate or overlap or combine, and thereby determine the texture of the whole."[8]

M: If, however, "veiled" means "illusory" or "inaccessible to concepts," then Buddhism would be in agreement with d'Espagnat.

T: I don't think d'Espagnat would call his "veiled reality" illusory. To his mind, it's a reality that escapes our perceptions and measuring apparatus. While I agree with you that interdependence must be the fundamental law, science can't describe it yet.

But even if scientists are having trouble grasping the fullness of interdependence, they are having no trouble finding a wide range of evidence for different kinds of interconnections in our world. For example, there is the cosmic interconnection of the Big Bang. We are all products of that primordial explosion. The hydrogen and helium atoms that make up 98 percent of the universe's ordinary matter were made during the first three minutes of its existence. The hydrogen in seawater and in our bodies all comes from that primordial soup. So we all have the same genealogy. As for the heavy elements that are needed for complexity and life, and which make up the other 2 percent of the universe's matter, they were produced by the nuclear alchemy in the center of the stars and the explosion of supernovas.

We are all made of stardust. As brothers of the wild beasts and cousins of the flowers in the fields, we all carry the history of the cosmos. Just by breathing, we are linked to all the other beings that have lived on the planet. For example, still today we are breathing in millions of atomic nuclei from the fire that burned Joan of Arc in 1431, and some of the molecules from Julius Caesar's dying breath. When a living organism dies and decays, its atoms are released back into the environment, and eventually become integrated into other organisms. Our bodies contain about a billion atoms that once belonged to the tree under which the Buddha attained enlightenment.

M: This also offers another way of looking at the EPR effect. Since all "particles"—whatever that might mean—were closely bound together in the singularity of the Big Bang (and perhaps during other Big Bangs), they must still be so now. Thus the natural condition for phenomena always has been and always will be global.

But in Buddhism, it isn't so much the molecular connections that matter—for they have little effect on our happiness or suffering—but rather the fact that all sentient beings, with whom we are *all* related through interdependence, wish to be happy and to escape suffering.

T: Yet another kind of interconnection discovered by science is that we're all linked together genetically. We all descend from *Homo habilis,* who appeared in Africa about 1,800,000 years ago, regardless of our race or skin color. As a child of the stars, humanity perhaps experienced a feeling of cosmic affiliation most intensely when we saw for the first time those stirring pictures from the space missions of our blue planet floating, so beautiful and yet so fragile, in the immense darkness of space. This global view reminds us that we are all responsible for our Earth and must save it from the ecological disaster that we're inflicting on it. William Blake expressed the global nature of the cosmos beautifully in the following lines:

To see a World in a Grain of Sand
And a Heaven in a Wild Flower,
Hold infinity in the palm of your hand
And Eternity in an hour.[9]

The entire universe is indeed contained in a grain of sand, because the explanation of the simplest phenomena brings in the history of the entire universe.

M: Those Blake lines remind me of one of the quatrains of a sutra by the Buddha:

As in one atom,
So in all atoms,
All worlds enter therein—
So inconceivable is it.[10]

Buddhist writings also say that the Buddha knows at all times the nature and the multifariousness of the universe's phenomena, in both space and time, as clearly as if he were holding them in the palm of his hand, and that he can transform an instant into an eternity, or an eternity into an instant. I can't help wondering if William Blake had read these texts, or whether inspiration passed down over the ages! If you consider these thoughts carefully, then you will see that the Buddha's omniscience corresponds exactly to a global perception. There's no need to see the Buddha as a god. It's enough to know that enlightenment embraces everything and knows at each instant the number and the nature of things. It is this global view that permits omniscience. The Indian Buddhist philosopher and poet Asvaghosa wrote, "As a result of deep concentration, one realizes the oneness of the expanse of reality."[11] On the other hand, fundamental ignorance results in fragmentation, and hence a limitation, of knowledge. We can perceive only certain aspects of reality, and fail to see its true nature.

T: The infinity of worlds makes me think of the other intelligent life-forms that probably exist in the cosmos. The observable universe contains several hundred billion galaxies, each having several hundred billion stars. If, like our sun, most of those stars have ten or so planets orbiting them, then we arrive at a total of several hundred billion trillion planets. It seems absurd that among such a huge number, our planet should be the only one to house conscious life. The existence of extraterrestrial civilizations raises interesting theological questions. For instance, Christianity says that God sent his Son, Jesus Christ, to Earth to save humankind. So are there a multitude of Jesus Christs visiting each planet that has conscious life in order to save the beings that have evolved there?

M: Buddhism talks of billions of different worlds, where different forms of beings live. It is said that most of these worlds have a Buddha who teaches the beings how to reach enlightenment. A Buddha doesn't save souls as one would throw a stone at a mountaintop. He gives them the means to identify the cause of their suffering and deliver themselves from it, and eventually to achieve the ultimate wisdom and bliss of enlightenment.

T: The Italian philosopher Giordano Bruno had already raised these questions at the end of the sixteenth century, when he suggested that the universe was infinite and contained an infinite number of worlds with an infinite variety of life-forms. He paid for such temerity with his life, for the Church condemned him to be burned at the stake four centuries ago, in 1600. It's fascinating to see that Buddhism was asking this sort of question more than two thousand years ago . . .

M: It is also said that on each blade of grass and each grain of dust, in each atom and in each pore of the Buddha's skin, there is an infinite number of worlds, but that it is not necessary for these worlds to shrink, or for the pores to grow larger. In other words, each element includes all of the others through interdependence without having to change its dimensions.

T: What a striking image! During our conversations I've greatly admired how Buddhism manages to use poetic images to express concepts that are often difficult, run against common sense, and can't be expressed in everyday language. According to Buddhism, does the world exist when it's not being perceived by a consciousness?

M: Of course, the world around us doesn't vanish when we are no longer conscious of it. But this is a false question because, to begin with, consciousness exists and is thus an active part of interdependence and, second, it would be impossible to imagine or describe what reality would be like if there were no consciousness. Thus, this position is neither nihilistic nor idealistic, in that it doesn't deny conventional reality. But neither is it realistic or materialistic, given that a reality existing only by its own means is meaningless for us. This is what the Buddha calls the Middle Way. In the words of a Tibetan commentator:

> *Two sticks which, when rubbed together, produce a fire, are themselves burned up in the blaze. Just so, the dense wood of all conceptual bearings of both existence and nonexistence will be totally consumed by the fires of wisdom of ascertaining that all phenomena lack true existence. To abide in that primal wisdom in which all concepts have subsided—this is indeed the Great* Madhyamaka, *the Great Middle Way, free from all assertions.*[12]

This is summed up by Nagarjuna in these verses from his major work, *The Fundamental Treatise on the Middle Way*:[13]

> *The words "There is," means clinging to eternal substance,*
> *"There is not" connotes the view of nihilism.*
> *Thus in neither "is" nor "is not"*
> *Is the dwelling place of those who know.*

In the *Sutra Requested by Sagaramati,*[14] the Buddha said:

The wise have understood interdependent origination,
They do not rely on extremist views.
They know that things have causes and conditions,
And that nothing is without cause or condition.

And Nagarjuna[15] went on:

That which arises dependent on something
Is not in the least that thing,
Neither is it different from it.
Therefore it is neither permanent nor nothing.

According to the Buddha,[16] the ultimate nature of phenomena is thus a union between appearances and emptiness:

Know that all phenomena
Are like reflections appearing
In a very clear mirror,
Devoid of inherent existence.

MIRAGES OF REALITY 5

ON THE EXISTENCE OF ELEMENTARY PARTICLES

Why is Buddhism interested in the science of elementary particles, given that studying them does not apparently have any particular effect on our daily lives? Well, if we ask questions about whether the world around us has a solid existence, it is important to study the nature of what are supposed to be its basic "building blocks." Buddhism is not alone in raising doubts about the "reality" of phenomena. The dominant explanation of quantum physics, known as the Copenhagen Interpretation, also suggests that atoms are not "things," but are "observable phenomena." This is a fascinating topic, because it places the human mind, or human perception, in the midst of what we call "matter" and "objective reality." If doubts can be raised regarding their "solidity," then many other conceptual barriers will fall down as a result.

MATTHIEU: Let's examine more closely the radical understanding in quantum physics of the dual nature of particles. The EPR experiment was based on the dual nature of photons of light. In the ordinary world, light is what allows us to see shapes and colors. If you sit in the sun with your eyes closed, then light is felt as heat. For a physicist who captures it with his instruments, light is a mathematical function, a set of numbers and equations. Each approach leads to a different description. Where does reality lie? Is it not more accurate to say that we're dealing with a set of interactions that create various transient phenomena, and behind this flow of endless transformations, we have no reason to postulate the existence of an intrinsic reality?

THUAN: Quantum physics agrees that light has no intrinsic reality, because it is neither exclusively a wave nor a particle. Instead, it can appear as either, depending on the circumstances. This does not mean that we don't understand a good deal about the ways that light interacts with the world, or that the different properties of light aren't real.

To consider your examples, colors can be explained by the particle aspect of light. Your monk's robes look red and yellow because the atoms they contain absorb blue and green, but reflect yellow and red. The photons reflected by your robes enter our eyes with an energy and frequency that create this impression of seeing red and yellow. If your robes simply reflected sunlight, without altering it, then they would appear as white as the sun. As for the heat of the sun, it too can be explained in terms of "grains of light." Each "grain" contains energy, and on a sunny day, countless grains of light collide with our skin, relaying their energy, which is transformed into heat.

Meanwhile, for the physicist, who sees light either as a wave or as a particle, depending on his measuring apparatus, if light appears as a wave, then its wavelength and frequency can be measured. If it appears as a particle, then its energy can be evaluated.

M: All we're doing here is describing some of light's observable properties,[1] but we are not really grappling with the question of light's ultimate nature. A photon is *never* simultaneously a wave and a particle. Almost 1,300 years ago, Shantideva wrote, "What arises through the meeting of conditions, and ceases to exist when these are lacking, is artificial like the mirror image. How can true existence be ascribed to it?"[2]

We must dig deeper to answer the question about the intrinsic reality of these particles. Can we say that we know light's *true nature*? Does such a nature in fact exist? Are not the "characteristics" of the particle simply the ways we perceive a transient phenomenon? Is an electron defined by its charge, its spin, or its mass? Is the entire set of its properties equivalent to the electron? In fact, are such properties intrinsic to the electron, or do they appear only in conjunction with the rest of the world, including ourselves?[3]

One of Buddhism's classic questions is, "Does a particle possess its properties in the same way that a farmer possesses a cow, or in the way that we possess our bodies?" If the former is true, then this would mean that the electron is distinct from its properties. If the latter is right, then the properties are part of the electron, because if we say that it possesses them, then this would mean that we had two bodies, the one we are and the one we possess. If the electron is each of its properties, then there would be as many electrons as there are properties. In this case, the electron entity would become multiple.

Building on these observations, I would like to transpose one of the arguments of the Buddhist philosopher Chandrakirti, who lived in India in the eighth century and taught at the famous Buddhist university of Nalanda, where up to twenty thousand students studied. He was one of the foremost philosophers of the Middle Way, and brilliantly deconstructed the notion of a "truly existing reality," which we would now call "material realism." In his original argument, the example of a chariot is used; I will replace that with the example of an electron. The whole argument is highly detailed,[4] but the gist of the argument is to explain that the electron doesn't really exist as a separate entity because (1) an electron is not its properties, for these are multiple and the entity "electron" would thus become multiple; (2) it is not something intrinsically different from its properties, because if it was so, it could then be perceived separately from its properties; (3) it is not the foundation of its properties; (4) its properties do not make up its foundation; (5) it is not the true owner of its properties; (6) it is not simply the sum of its properties; and (7) it is not the form of its properties. If it is neither equivalent to its properties nor separate from them, the entity "electron" is a mental label with a purely conventional existence. Alan Wallace wrote, "Human beings define the objects and events of the world that we experience. Those things do not exist intrinsically, or absolutely, as we define or conceive of them. They do not exist intrinsically at all. But this is not to say that they do not exist. The entities that we identify exist in relation to us, and they perform the function that we attribute to them. But their very existence, as we define them, is dependent upon our verbal and conceptual designations."[5]

T: I agree with this view because quantum theory backs it up. The discovery of light's dual nature was certainly a great surprise for physicists. But what's even stranger is that matter has exactly the same duality. What we call an electron, or any other of the elementary particles, can also appear as a wave. Thus the particle and wave aspects cannot be dissociated; rather they complement one another. This is what Niels Bohr called the "principle of complementarity." He saw this complementarity as the inevitable result of the interaction between a phenomenon and the apparatus used to measure it. According to him, it isn't so much reality that is dual, but the results of experimental interactions.[6]

The act of observing also introduces quantum fuzziness. This is expressed in Heisenberg's uncertainty principle, which tells us that it is impossible to define precisely at the same time an electron's position and its speed. To determine the position of an electron, we have to shed light on it. But the photons in the light relay their energy to the electron in this process, and the higher the energy, the more they disturb its movement. We are thus up against a dilemma: the more we decrease the uncertainty of the electron's position by shining light on it, so that we can see it, the more we increase the uncertainty of its movement. On the other hand, if we use only low-energy light, we don't disturb its movement much, but we increase the uncertainty of its position. The act of determining the one aspect of the electron eliminates the possibility of determining the other. Thus, talk of an "objective" reality without any observer is meaningless, because it can never be perceived. All we can do is capture a subjective aspect of an electron, depending on the observer and the apparatus used. The form that this reality then takes is inextricably bound up with our presence. We are no longer passive spectators faced with a tumult of atoms, but full participants.

M: But this still tells us nothing about the ultimate reality of this particle—if such a reality exists. Neither the particle nor the wave, nor, for that matter, any other entity, exists inherently. For example, I suppose that we can't affirm that the particle existed before it was observed?

T: Before measurement, all we can talk about is a wave of probability.

M: If when we say "particle" we mean something with an intrinsic or even permanent reality, and if it didn't exist before it was observed, nothing could bring it to life. How could an entity that contains all the qualities we usually attribute to a particle abruptly pass from nothingness to existence? When a particle appears, either it does not exist independently as an entity, or it has been created *ex nihilo.*

T: And yet before, there was a wave. There was something, not a complete vacuum!

M: Buddhism doesn't talk about a complete vacuum—that would be nihilistic—but "lack of intrinsic existence." It is for this reason that, depending on the circumstances and on the experimental technique, an unreal phenomenon can appear to be either particle or a wave.

T: Our debate here is precisely the one that went on between Einstein and the originators of the Copenhagen Interpretation of Quantum Physics, Niels Bohr, Werner Heisenberg, and Wolfgang Pauli. The interpretation is given this name because the institute run by Bohr, where Heisenberg and Pauli were frequent visitors, was in Copenhagen. In simple terms, it says that "atoms form a world of potentials and possibilities, rather than of things and facts." According to Heisenberg, "in quantum physics, the notion of a trajectory does not even exist."[7] This view could not be further from Einstein's dogmatic realism.

This is how Heisenberg summed up Einstein's counterargument: "This interpretation does not describe what actually happens independently or in between the observations. But something must happen, this we cannot doubt. . . . The physicist must postulate in his science that he is studying a world which he himself has not made and which would be present, essentially unchanged, if he were not there." We could call this position of Einstein's one of material realism.

Heisenberg's response to this objection of Einstein's is complex, but I think it is important to offer in his own words:

> *It is easily seen that what this criticism demands is again the old materialistic ontology. But what can the answer from the point of view of the Copenhagen interpretation be? . . . The demand to "describe" what "happens" in the quantum-theoretical process between two successive observations is a contradiction* in adjecto, *since the word "describe" refers to the use of classical concepts, while these concepts cannot be applied in the space between the observations. . . . The ontology of materialism rested upon the illusion that the kind of existence, the direct "actuality" of the world around us, can be extrapolated into the atomic range. This extrapolation is impossible, however.*[8]

M: A Buddhist philosopher would be in complete agreement with this answer.

T: Personally, I also agree with Heisenberg. As I've already said, quantum mechanics has always been confirmed by experimentation and has never been caught out. Einstein got it wrong, and his material realism cannot be defended. According to Bohr and Heisenberg, when we speak of atoms and electrons, we shouldn't see them as real entities, with well-defined properties such as speed and position, tracing out equally well-defined trajectories. The "atom" concept is simply an image that helps physicists put together diverse observations of the particle world into a coherent and logical scheme. Bohr also spoke of the impossibility of going beyond the results of experiments and measurements: "In our description of nature the purpose is not to disclose the real essence of phenomena but only to track down, so far as possible, relations between the manifold aspects of our experience."[9]

M: His viewpoint is similar to that of my former teacher François Jacob, who said, "It thus seems clear that the physicists' description of atoms is not the exact and unchanging reflection of a revealed truth. It is a model, an abstraction, the result of centuries of effort focused by physicists on a small number of phenomena in order to construct a coherent picture of the world. The description of the atom is as much a creation as it is a discovery."[10] But this doesn't stop most people from imagining atoms as little balls they could pick up if they had tools that were small enough.

T: Schrödinger warned us against such a materialistic view of atoms and their constituents: "It is better not to view a particle as a permanent entity, but rather as an instantaneous event. Sometimes these events link together to create the illusion of permanent entities."[11]

M: The ring of light created by a rotating flashlight isn't an "object." The world of phenomena is made up of events that can't remain stable from one moment to the next. If they did, they'd stay frozen forever. Since such moments are transient, they have no duration, and the events concerned cannot exist independently. So we cannot assume that, one day, we'll know all of the characteristics of the event "particle." It appears to us in different forms because of interdependence, which is synonymous with the "absence of intrinsic being."

The essential point is that a phenomenon's characteristics do not belong to it intrinsically. For instance, when we say that mass can be converted into energy, this comes down to saying that mass isn't a property that we can always associate with the "particle event."

T: That's right. As with light, the nature of matter isn't immutable. Energy can be converted into matter. This is often done in particle accelerators. Energy can come from mass (as in Einstein's famous equation $E = mc^2$) or from movement. In the latter case, this means that an object's property can be converted into an object. Inversely, matter can be converted into energy—this is what makes the sun shine, for example. By converting a tiny fraction of its mass of hydrogen (0.7 percent) into light (photons), our star allows life to exist on earth.

M: This implies that neither of these mutually exclusive properties really constitutes what we call a photon. If they did, then the properties should *always* be present. What would we make of an animal that looked like a cat from one side and a dog from the other?

So reality doesn't lie in the solid concepts we attach to things. Phenomena can appear without having any underlying, intrinsic reality. We must transcend our conceptual limitations, which make us think that things must either exist intrinsically, or not exist at all. There is a middle way, which can be expressed as a dream or mirage. A phenomenon can still function even if it is just an illusion. A reflection in a mirror can appear and disappear, be transformed in various ways, and communicate different sorts of information, even if nothing "came into existence" in the mirror.

T: A Platonist would tell you that the mirror world is simply a reflection of the real world. However, it is certainly true that the particle aspect is no more basic than the wave aspect. So we must say that neither light nor matter have intrinsic, immutable properties. Such properties depend on the observer and the apparatus. So, given that they are impermanent, they can be considered to be "illusory."

M: Is physics prepared to admit that an electron is merely a product of relationships and has *no* fundamental reality?

T: If, when you say "relationships," you mean the interactions between the observer and the observed and the interactions and transformations of elementary particles (for example, a proton and an electron come together to form a neutron or a neutrino), and the interaction between matter and light, then I would certainly agree.

M: By "relationships," I don't mean interactions between distinct, intrinsically existing objects, but a network of infinite relationships that condition each other mutually. A phenomenon's apparent properties derive from the complete set of phenomena, consciousness included.

T: That reminds me of what Heisenberg said: "The world thus appears as a complicated tissue of events, in which connections of different kinds alternate or overlap or combine, and thereby determine the texture of the whole." [12]

M: The word he used is *events* and not objective entities. A particle seems to be distinct from global phenomena simply because we are studying it and have isolated it by means of an experimental protocol and our concepts. But no given part of this global world of phenomena can possess fundamental characteristics. . . .

T: Your remarks bring to mind a theory that was fashionable in the 1960s, which stated that there were *no* elementary particles. Each particle would thus be composed of all the others, and there would be a little of every particle in each: *a* is made up of *b* and *c*, *b* of *a* and *c*, and *c* of *a* and *b*. The American quantum physicist Henry Stapp wrote: "An elementary particle is not an independently existing unanalyzable entity. It is, in essence, a set of relationships that reach toward other things."[13] This "bootstrap" theory is no longer in favor because of a lack of experimental proof. A scale of smaller and smaller particles—molecules, atoms, electrons and nuclei, protons and neutrons, quarks—seems to provide a better description of our observations of atomic and subatomic phenomena.

M: And yet some philosophers of science, such as Bernard d'Espagnat and Michel Bitbol, criticize this picture for being a somewhat vulgar generalization of our gross perceptions. Bitbol affirms that quantum events can be explained "just as well using a model of substitution in which there are *no* body-like elements."[14] This chimes with what Schrödinger said: "Modern atomic theory has been plunged into a crisis."[15] Bitbol goes on to explain: "We must not forget that, in general quantum terms, the possibility of individualizing objects at the atomic level is restricted to certain well-defined experimental conditions, and that it vanishes completely when these conditions are no longer fulfilled." He also quotes Quine, who wondered if quantum theories haven't imposed an about-face on physics in which "theories can take yet more drastic turns: such not merely as to threaten a cherished ontology of elementary particles, but to threaten the very sense of the ontological questions, the question 'what there is.'"[16]

The physicist Laurent Nottale remarked:

> *Some philosophers have gone farther and concluded that nothing, including matter and mind, intrinsically exists. If we trace the*

history of this line of thought back, it seems to have been first formulated in Oriental thought by Siddharta Gautama over two thousand five hundred years ago. There is no nihilism in this concept, no denial of reality or existence, but rather a profound view of the very nature of existence. If things do not exist in absolute terms, but do nevertheless exist, then their nature must be sought in the relationships that bring them together. Only these relationships between objects exist, and not the objects themselves. Objects are relationships. . . . Will the physics of the future succeed in making an equation of what is now a purely philosophical vision?[17]

T: Only time will provide the answer to that question. But certainly, if physics is to move forward in that pursuit, it must answer some other tricky questions about quantum phenomena first. For example, macroscopic objects, such as a table, or this book, are made up of particles governed by quantum uncertainty. So why can't this book suddenly vanish and reappear outside in the garden? The laws of quantum mechanics state that such an event is possible in principle, but it is so improbable that it could happen only if we waited for all eternity. Why is it so unlikely? The reason is that macroscopic objects consist of such a huge number of atoms (a book contains about 10^{+25} and the Earth about 10^{+50}) that the effects of chance cancel each other out. The probability of finding this book in the garden is infinitely small, because a large number of atoms implies a large mass and so high inertia. Ordinary objects are not really disturbed when observed under light, because the energy relayed by the light is negligible. Thus the speed of such objects can be accurately measured along with their position. Quantum uncertainty is eliminated. But where does the borderline lie between the microcosm, ruled by quantum uncertainty, and the macrocosm, where it fades away? Physicists are still unable to define this frontier, even though they are daily rolling back the limits of the quantum world. A molecule of fullerene, made up of sixty carbon atoms, is the largest object that has so far been seen to behave in a wavelike manner.[18]

M: Perhaps there is no frontier. Uncertainty doesn't vanish, it simply becomes imperceptible in a macroscopic environment. In the same way, we don't perceive the effects of space-time relativity because our movements in relation to others are nowhere near the speed of light. However, relativity continues to apply: a bicycle becomes smaller when it starts to roll, but this change is so tiny that a stationary observer can't see it.

The *possibility,* even if it is infinitely small, that this book could vanish and reappear in the garden shows that there is no *fundamental* difference between the microcosm and the macrocosm. Even if in our daily lives we are on a level where uncertainty can't be perceived, this doesn't refute the quantum nature of the world. Henry Stapp, the quantum theorist you mentioned, wrote, "The important thing about Bell's theorem is that it puts the dilemma posed by quantum phenomena clearly into the realm of macroscopic phenomena. . . . [It] shows that our ordinary ideas about the world are somehow profoundly deficient even on the macroscopic level."[19]

The biggest problem for realists is to reconcile the discoveries of quantum physics with daily reality in the macrocosm. Physicists keep shifting from one to the other. One moment they talk of particles and actual objects, the next of complementarity and nonlocalization. They should use the conclusions of quantum mechanics to transform their vision of the world.

Why should there be a line in the sand between the macrocosm and the microcosm that constitutes it? The former is simply an extension of the latter. What emerges when the microcosm becomes the macrocosm? A structure—that is to say, a set of relationships resulting in *functions* possessing a certain *continuity* and capable of *transforming* phenomena. However, these functions do not give any more reality to this structure or to its parts than particles have. If particles aren't "things," then daily reality doesn't consist of "things" either, no matter how it might look.

T: But that raises questions about how the macrocosm works. We're surrounded by macroscopic objects, with well-defined speeds and positions, which are not subject to Heisenberg's uncertainty principle, and do not have the wave/particle duality of the atomic and subatomic worlds. Chance is neutralized in the macroscopic world. The objects it contains can't be everywhere at the same time, like a wave. As I've already said, I'm hardly likely to find your watch in my pocket (unless I become a sneak thief!) or to see our Moon start orbiting around Mars.

M: But their solidity and dependability is a mere illusion. According to Buddhism, the perception of solid reality is caused by the momentary stabilization of a network of relationships. A dream that lasts a century is no more real than a dream that lasts only one minute.

The way we describe reality is conditioned by the fact that our daily experience allows us to see only the macroscopic level, where stability is highest. It is quite likely that if we had the microscopic level before our eyes, then we wouldn't consider the exterior world to be solid. Our perceptions of this world depend entirely on how we stand in relation to it. According to some physicists, such as Laurent Nottale, the apparent incompatibility between classic mechanics and quantum mechanics is simply a question of "scale relativity."[20]

Let's take the "tent" example used in Buddhist analysis. If we dismantle a tent, by separating its cloth, poles, and ropes, then the tent no longer exists. But the parts still exist. So let's tear up the cloth. There are then threads that we can reduce to fibers, then to molecules, then to atoms, and finally to particles whose mass is equivalent to intangible energy. This transition from a tent to the unreality of particles or, the other way around, from particles to a tent, contains no discontinuity that could justify a distinction between the microcosm and the macrocosm. So why do we see the tent as having a greater degree of reality? Because we are approximating and not investigating thoroughly. As the Buddhist texts say, "Because of a lack of critical inquiry, we eagerly accept that things are as they seem."[21] Quantity changes nothing. A million particles are no more real than just one. The nonreality of particles is proof enough of the nonreality of macroscopic phenomena. As said by Nagarjuna in the *Ratnamala,* "If the seed of something is unreal, how can the sprout be real?"

In the same text, he observed, "The farther we are from things, the more real they seem. The closer we get to them, the more they elude our grasp; they are like a mirage, unreal, intangible." If an elementary particle is neither cloth nor pole, and neither heat nor color, then neither is it "me" or "you." It thus escapes from the mental constructs that cause our mismatch with the world and, hence, our suffering. This is what Shantideva was saying when he spoke of knowledge that transcends discursive thought: "When real and non-real both are absent from before the mind, nothing else remains for the mind to do but rest in perfect peace, from concepts free."[22]

T: I can see no real contradiction between the ideas of science and Buddhism when it comes to the reality of elementary particles. The notion of an elementary particle is of course similar to the notion of an atom. Doesn't Buddhism talk of atoms?

M: Several centuries before Christ, at the time of the great Greek philosophers, Buddhism undertook a logical analysis of the notion of the atom—that is to say, etymologically, something that is "indivisible." But first, perhaps you should remind us of the ideas of Leucippus and Democritus, the first people to come up with this notion.

T: The concept of the atom is one of the foundation stones of scientific history. The American physicist Richard Feynman even declared that if all scientific knowledge vanished during some terrible upheaval, then the sole concept that should be preserved for future generations is that "all things are made of atoms—little particles that move around in perpetual motion."[23]

This idea dates back to the sixth century B.C.E., when two Greek philosophers, Leucippus and Democritus, introduced the revolutionary notion that all matter is made up of eternal, indivisible particles, which they called "atoms" (from the Greek *atomos*—"something indivisible"). Because of a lack of experimental proof, this idea remained a mere philosophical proposition for twenty-one centuries, and was eclipsed by the famous Aristotelian quartet of elements: earth, air, fire, and water. Only around 1600 did the atom idea resurface. In 1869 the Russian chemist Dmitri Mendeleyev had the brilliant idea of organizing the elements according to their atomic weight. As if by magic, elements having the same chemical properties came together in groups of seven, thus forming what we now call the periodic table of the elements. Such an arrangement can be explained only if each chemical element consists of just one specific type of atom. When Mendeleyev had drawn up his table, only sixty-three elements had been discovered. He was so sure that his table was right that he had no qualms about leaving blank spaces. And history was to prove him right. The blank spaces were subsequently filled in as new elements were discovered.

M: But the idea that matter consists of eternal, indivisible particles must already have existed in India at the same time as the early Greek philosophers were expressing it, because Buddhist thinkers went out of their way to refute it. They said that if a particle is indivisible, then it must be a dimensionless point.

T: That was presumably an image, given that they could hardly have known about the mathematical concept of points.

M: The concept is implied in their argument, which runs as follows: Let us suppose that matter does consist of indivisible particles. It would then be necessary for these particles to become associated. Can two supposedly indivisible particles come into contact? Don't forget that this was a thought experiment.

So let's suppose that these two indivisible particles come into contact. Do their entire surfaces touch simultaneously or gradually? If the latter is true, then the western side of a particle, for example, would first touch the eastern side of a second particle. But if these particles have a west side and an east side, then they are made up of parts and so cannot be indivisible. But if we reply by saying that they have neither parts nor sides, then they also lose their dimensions. Thus the only way they can come into contact is to fuse together. Now, if two particles can fuse together, then why not three? A mountain or the entire universe could fuse with just one particle. Everyday reality could thus neither crystallize nor unfold. This reasoning *ad absurdum* led Buddhist philosophers to claim that the universe cannot consist of individual, indivisible particles.

T: I could answer by saying that the particles don't need to come into contact in order to form matter.

M: In that case, the philosophers went on, there must be an empty space between the particles and, given that they have no dimension, an infinite number of particles and the entire universe, for that matter, could come between two particles.

T: Such sophisticated thinking is all the more astonishing given that it was apparently developed independently from the Greeks. Was it before or after them?

M: Around the same period, in the sixth century B.C.E. This concept was then further elaborated in a large number of philosophical works, issuing from debates between Buddhism and Hinduism and between different schools of Buddhism until the seventh century C.E. The Buddha taught according to the faculties and attitudes of his audience. To some he said that matter existed, to others that it was "unreal but apparent." As ever, we immediately come to the therapeutic aspect of the investigation, which aims to free us from the suffering caused by our attachment to reality. It is thus a step toward enlightenment.

T: If the concept of the atom emerged at about the same time in Greece and in India, do we know if there were contacts between these two schools of thought? Who are India's Leucippus and Democritus?

M: The Buddha himself, about 2,500 years ago, then his main followers such as Nagarjuna and Aryadeva (second century C.E.), Vasubandhu (fourth century C.E.), and Chandrakirti (eighth century C.E.) all discussed this issue. Before them, Hinduism said that matter was composed of microscopic elements arranged in a contiguous way. Others, including some Buddhists who were more materialist than those I've just cited, thought that atoms were arranged like grains in a cup, so that only certain points touched. In their opinion, matter looks continuous to us simply because we can't examine it closely enough, just as a field looks like a large patch of green from a distance, whereas it in fact consists of a multitude of distinct blades of grass.

T: The idea that matter consists of contiguous atoms is of course in keeping with the claims of modern physics. We now know that atoms are almost entirely empty. The nucleus accounts for 99.9 percent of an atom's mass but takes up only a thousandth of a trillionth of its volume. The rest is occupied by a cloud of swirling electrons. Matter looks continuous to us, because our eyes can't see the atomic level, at about one hundredth of a millionth of an inch.

M: As for exchanges between Greek and Buddhist philosophers, they certainly took place, as can be seen in the discussions between Menander, the king of Bactria, whose culture was Greek, and a Buddhist monk called Nagasena, between 163 and 115 B.C.E.[24] But it's difficult to gauge the importance of such influences. It should also be said that Democritus's concept of the atom was more rudimentary. He speaks, for example, of "hooked atoms" that come together according to their various affinities. Dignaga, the great fifth-century C.E. Buddhist logician, would no doubt have objected that if they had hooks, then they had parts, and so were not indivisible!

T: And so the philosophical idea of the atom gradually became a scientific concept. But by the middle of the nineteenth century, science still hadn't proved Leucippus and Democritus's fundamental idea that matter was made up of indivisible elementary particles. In fact, Mendeleyev's work rather suggested the contrary, given that if chemical elements could be organized in a periodic table according to their atomic weight, then atoms must have different degrees of complexity, with the heaviest ones being the most complex. In that case, atoms must be made up of even smaller particles. This was to be confirmed by experiment. When studying electric discharges in gas, Joseph Thomson, a British physicist, discovered in 1897 that each atom contained particles carrying a negative electric charge, and the number of them equaled the atomic weight. This new particle was called an electron, which means "amber" in Greek, because the Greeks had discovered that amber, when rubbed with wool, had a mysterious power of attraction.

But the most astonishing results were obtained by another British scientist, Ernest Rutherford, in 1910. By bombarding thin sheets of gold leaf with highly energetic particles, he noticed that the vast majority of these particles crossed the gold leaf as though it didn't exist, while a tiny fraction (0.01 percent) of them were reflected and returned to their starting point. It was as if a rifle bullet had been sent back by a sheet of paper! Before Rutherford's experiment, physicists thought that atoms occupied almost all the space inside a solid object, like apples in a barrel, with only a tiny gap between them. If that was the case, then none of the particles Rutherford sent toward the gold leaf should have been knocked back. The explanation must be that atoms had a hard, dense nucleus capable of reflecting particles. This nucleus must occupy a tiny space in comparison with the total volume of the atom, since the majority of the projectiles missed it and continued their journey unaffected. We now know that an atom's nucleus occupies the same space as a grain of rice in a football stadium. Thus, all of the matter around us, that sofa, the chair, the walls and so on, is almost totally empty. The only reason we can't walk through walls is that atoms are linked together by the electromagnetic force.

And so the concept of emptiness emerges once more, but not the primordial vacuum that gave birth to the universe and the matter it contains, but the emptiness of atoms. The surprising fact is that Leucippus and Democritus also mentioned the vacuum when introducing their notion of the atom.

M: But this void is completely different from modern physics' energy-filled vacuum and Buddhism's emptiness, which isn't the absence of "something," but the absence of "intrinsic nature."

T: That's right. If atomic theory generally goes about describing only one sort of reality (that of atoms), Democritus and Leucippus presented it in terms of a duality: corpuscules and the void, empty of all matter, are united in a complementary and associated manner within the same reality. Simplicius, the Greek historian, described their vision in the following terms: "Leucippus and Democritus hold that the worlds, which are in unlimited numbers and occupy the unlimited void, are formed by an unlimited number of atoms."[25] Greek philosophers thought that the void surrounded the atoms, whereas Rutherford discovered that it was inside.

M: In fact, it comes down to much the same thing: we just have to drop the word "atom" and say "nucleus" instead. But an atom's nucleus isn't indivisible.

T: Exactly. We now know that atomic nuclei consist of protons and neutrons, held together by the strong nuclear force. Protons and neutrons are extremely similar, except for their electric charges. The proton's charge is positive, equal to and opposite to the electron's. As its name implies, the neutron has no electric charge. Their masses are almost identical: about two thousand times that of the electron. Neutrons give matter its stability and see to it that the objects around us aren't constantly disintegrating. If atomic nuclei were made up only of protons, they'd break up, because particles with the same charge repulse one another. The books on the shelf behind you, the cup of tea on the table, and the roses in the garden would immediately fall to pieces.

To return to Buddhism's refutation of the existence of indivisible particles, it agrees with certain notions or discoveries in subatomic physics, but it contradicts others. The "standard theory," which for the moment best explains the properties of particles on the subatomic level, says that indivisible[26] particles called quarks are the building blocks of other particles. Their inventor, an American physicist called Murray Gell-Mann, so christened them in 1963 because he liked the sound of the sentence "Three quarks for Muster Mark" in James Joyce's *Finnegans Wake.* As for "Muster Mark," three quarks are needed to form a proton or a neutron. The electric charges of quarks are fractional ($+/-{}^{1}/_{3}$ or $+/-{}^{2}/_{3}$) because their sum must equal the charge of a proton (+1) or a neutron (0).

Quark theory is generally accepted because it successfully accounts for the properties of the hundreds of known particles. The vast majority of them live for only a minuscule fraction of a second. They don't appear in the matter around us and are born during the collisions between particles in accelerators. In the same way that Mendeleyev ordered the chemical elements in his periodic table, Gell-Mann managed to use quarks to explain the "zoo" of particles that was being discovered by physicists during the 1960s.

M: Thus we have apparently gone back to a reifying vision of particles, despite the wave/particle complementarity of the Copenhagen Interpretation of Quantum Mechanics.

T: That's right, we now have to see whether quarks really exist, or whether they're just theoretical entities that have been invented to sort out the world of particles. The hunting of the quark was in full swing in the 1960s, but we have never been able to identify a particle with a fractional charge. In 1968, physicists used the two-mile-long linear accelerator at Stanford University to shoot beams of extremely energetic particles toward protons, in the hope of breaking them up and thus freeing some quarks. But they failed. However, the way the particles ricocheted off the protons seemed to show that these are structured and made up of three pointlike parts.

Why is it impossible to observe independent quarks? We can imagine that the strong nuclear forces that bind together the three quarks inside the proton act like strings. Our intuition says that if we pull harder and harder on a string, then it will break, thus freeing the quarks. But this is to forget the fact that by pulling on the string, we add energy to it, and that this energy is liberated when it breaks. Because of the equivalence between energy and matter, the energy that has been freed creates a quark/antiquark pair (the antiquark has the same properties as a quark, but with the opposite charge). Two events then occur: the new quark that has been produced immediately replaces the freed quark to reconstitute the proton; and the new antiquark combines with the freed quark and together they form a new particle called a meson. In the end, we haven't freed a quark. All we've done is to create a meson. Quarks can never exist independently. We'll never see one. Trying to extract the quarks from a proton is like trying to isolate one of the poles of a magnet: if we cut it in half, we don't obtain isolated poles, but two smaller magnets, each having its own north and south poles.

M: If a quark interacts with an antiquark to form a meson, then it loses its identity. Can we then continue to see it as the basic building block of matter?

T: That's what Gell-Mann postulated, because it's unnecessary to say that quarks are made up of even smaller particles in order to understand how protons, neutrons, and hundreds of other particles behave and to organize them into a logical setup. This is Occam's razor once again. Scientific methodology consists in explaining the greatest number of facts with the smallest number of hypotheses. An additional hypothesis can be added only if some facts remain unexplained.

M: It's one thing to have no need of the idea that quarks are made up of smaller entities right now, and quite another to state that they are the intrinsically indivisible building blocks of matter. Mendeleyev's table has allowed us to explain the properties of the chemical elements using a simple, logical structure. But this didn't show that the atoms in these elements are indivisible. From what you've said, all we can claim is that quarks cannot be broken up by the energy that we are currently capable of throwing at them. But does this imply that quarks are really indivisible? Do they have dimensions?

T: Since we can never see one, we have no idea about their possible dimensions, except of course that they must be smaller than a proton, which measures just a ten-thousandth of a billionth of a centimeter (10^{-13}). As regards their indivisibility, that is of course postulated in the theory.

M: Such a description of atoms is, I suppose, extremely comforting for proponents of "realism," and it provides us with handy images that describe nature in terms of our view of the macrocosm. But it shouldn't be forgotten that quarks, too, are subject to the particle/wave duality. We thus return to our initial discussion about the reality of such particles.

According to Buddhism, this reifying standpoint, based on our ordinary perceptions and common sense, is at the heart of the mismatch between the nature of phenomena and how we perceive them. Numerous scientists and philosophers of science have said the same; their writings are full of warnings against overly simplistic images or generalizations of the atomic world. Heisenberg wrote that "atoms are not things," and that "the ontology of materialism rested upon the illusion that the kind of existence, the direct 'actuality' of the world around us, can be extrapolated into the atomic range. This extrapolation is impossible, however."[27]

Bitbol states that "atomism does not describe an ascent through phenomena toward an optimal explanation; it is a structure used to anticipate phenomena and to guide the experiments that help to define them"[28] and "fundamentally, atomism is a particular example of a wider tendency to found appearances on a universe of forms."[29]

I use the preceding quotes to show that even if it isn't the majority view, realism is sometimes criticized in modern scientific thought, just as it always has been by Buddhism. It's obvious that many of our contemporaries have problems with drawing all the logical conclusions from the Copenhagen Interpretation of Quantum Physics and the global nature of phenomena as shown in the EPR experiment. If that is the true nature of phenomena, then understanding it should radically alter the way we perceive gross reality—ourselves and the world around us.

This is Buddhism's basic approach. Not only is this a part of knowledge, but it is also an aid to personal transformation. The analysis that leads to an understanding of emptiness can, at first glance, appear highly intellectual. But the resulting comprehension frees us from our attachments and thus brings about radical changes in how we lead our lives.

LIKE A BOLT FROM THE BLUE 6

THE IMPERMANENCE AT THE HEART OF REALITY

The impermanence of phenomena is not only a subject for meditation and one that should incite us to make good use of the time we have left to live. Impermanence is also an essential part of our understanding of reality: it should determine both our vision of the basic nature of the world and our behavior. We must ask, are there or are there not permanent entities in the universe? If nothing is permanent, as Buddhist analysis shows, and physics seems to confirm, then how should this knowledge influence the way we live?

MATTHIEU: In Buddhism, the concept of interdependence is closely linked to the notion of the impermanence of phenomena. Impermanence is a correlate to interdependence. We distinguish in Buddhism between gross impermanence—such as the changing seasons, the erosion of mountains, the passage from youth to old age, or our varying emotions—and subtle impermanence, which takes place in the shortest conceivable period of time. At each infinitesimal moment, everything that seems to exist changes. Once we have recognized that change is inevitable and omnipresent, we can understand that the universe is not made up of solid, distinct entities, but of a dynamic flow of incessant interactions. We've seen how what we call "matter" should be seen in various ways—a wave or a particle, mass or energy—which can't coexist simultaneously. This seems to imply that nothing can be seen as being a permanent entity. What does science have to say about this question of the longevity of matter, of such particles as quarks, for example? Are quarks really as fundamental and as permanent as they are sometimes described as being?

THUAN: To answer this question, we should introduce the science of quarks a little more fully. In order to explain the mind-boggling variety of particles they've discovered, physicists have come up with several types of quark, which they classify poetically in terms of their "flavor" and "color." The first family includes quarks whose flavor is either "up" or "down," the second family includes "strange" or "charmed" quarks, and the third family contains "bottom" and "top" quarks. Moreover, for each quark flavor, there can be three colors called "yellow, red, and blue." So there are eighteen sorts of quark in all. Ordinary matter, consisting of the protons and neutrons in atomic nuclei that make up our bodies and the flowers in the field, is built from just the "up" and "down" flavors. The other varieties occur only in high-energy particle accelerators. To answer your question, quarks in the same family can, and regularly do, change flavor. An "up" quark, for example, can become a "down" quark, and vice versa. Quarks can also change families, but only if they change their electric charge.

M: So the state a quark is in is not permanent?

T: That's right. And changes in the states of quarks lead to changes in the protons and neutrons that they form. For example, if an "up" quark changes into a "down" quark, the proton becomes a neutron[1] and emits a positron, which is the antiparticle of the electron, and a neutrino. A neutrino is a particle with zero or insignificant mass that hardly interacts at all with ordinary matter. As I speak, hundreds of billions of neutrinos created at the beginning of the universe are streaming through our bodies every second.

M: So a quark is neither immutable in its nature nor eternal, and is therefore not really an "ultimate particle." However, don't some physicists think that certain particles would be eternal if left to their own devices (which in any case is a purely theoretical situation that could never naturally occur)?

T: Of the hundreds of known particles, physicists think that only a few would be immortal. Most of the types of particles that appear in accelerators are so unstable that they live for only a millionth of a second, or even less. If we liberate the neutron, which, along with protons, makes up the atomic nuclei, it survives for only about fifteen minutes when it is free, before spontaneously transforming itself into a proton, emitting an electron and an antineutrino in the process. But so long as it stays locked up inside a nucleus, it too is practically immortal. This is just as well for us, because otherwise our bodies would disintegrate in a quarter of an hour!

If left to their own devices—that is to say, if they weren't bombarded by other particles—only the electron, the photon, and the neutrino would live forever.

M: How can you prove that?

T: Of course, we can't measure eternity, but we can experimentally establish life spans that are so long that they're practically eternal. What physicists call the "standard model," the theory that currently best explains the world of particles, has no need of any hypothetical death for electrons, photons, and neutrinos.

As for the proton, recent theories estimate its life span as being a hundred million trillion times the age of the universe, which is 10^{+30} years. Of course, to check this, physicists can't wait 10^{+30} years to see a proton die. Instead we've devised a clever way of estimating proton life in much less time. Quantum mechanics states that it is theoretically possible for any given proton to disintegrate at any moment. If we could collect a huge number of protons somehow and watch them carefully, we might see one die. If a proton's life span is 10^{+30} years, then we simply need to bring together 10^{+30} protons in order to see one of them die per year. Even better, if we assembled 10^{+33} protons, then we would see several of them disintegrate every day. To achieve this, Japanese physicists have constructed a huge cylindrical observation tank buried one thousand meters underground in a mine in the town of Kamioka, Japan, which is named Super-Kamiokande (often called Super-K). The tank is forty-one meters in height and thirty-nine meters in diameter, and it's filled with fifty thousand tons of water. Water is a good source of protons, and the point is to monitor the water in the tank for signs of any protons decaying. It's located so far underground because this filters out cosmic rays, which might trigger reactions that mimic the signs of real proton decay. Observations began on April 1, 1996, but so far, no one has seen a single proton die. This means they live longer than the standard model predicted. Compared to a human life, 10^{+30} years plus is almost an eternity.

M: But it still isn't in fact eternal. Even if the electron and the neutrino are theoretically immortal and don't spontaneously disintegrate, they can still be transformed. To do this, they simply have to be bombarded with other particles containing enough energy.

T: True. So far, I've been talking about the unprovoked death of particles. But it's true that we can easily make them vanish by bombarding them, or making them interact with other particles. When a proton interacts with an electron, it becomes a neutron and emits a neutrino. Let's take a solar photon. When it interacts with matter—with the wood of this table, for instance—it loses part of its energy, which is converted into heat. Its nature has now changed. It can even lose all of its energy and thus disappear while warming up this table.

M: Disappear? So much for its immortality! So it's clear that all particles are impermanent and die sooner or later.

T: That's one way of putting it. The particles that make up matter are either unstable, and so break up spontaneously, or else stable, but then interaction with other particles makes them alter their nature or even disappear. However, I repeat that stable matter that is left alone will not change perceptibly. You'll not see this flowerpot vanish in front of your eyes.

M: But that's just a question of time and perception. The matter that you call stable is constantly changing in an imperceptible way, otherwise it wouldn't age.

T: True enough. Everything that is left to itself ages and decays. The varnish on the table loses its brightness, a badly maintained house falls into ruins. All forms of order, of substance, in the universe tend toward disintegration, or disorder, over time. This is the famous second law of thermodynamics, which states that the total quantity of disorder in the universe must always increase, or at least can never decrease.

M: This continuous kind of change is what we call subtle impermanence. If we didn't have this subtle impermanence, then things couldn't change now or in the future. If something could remain the same even for a moment, then it would be stuck in its present form for all eternity. Nothing could happen. The cause of an object's destruction lies in its birth. Thus impermanence is at the center of the process of causality. What matters is that *nothing,* no particles or any other entity in the universe, can be *absolutely* permanent. This is an essential point for the elucidation of phenomena.

The central question is, does reality consist of immutable elements that exist through their own agency? If the quark, which you consider to be matter's fundamental element (and if this is so, why are there eighteen different sorts?), was permanent, it would need no other cause than itself in order to exist. But, as Nagarjuna pointed out, "Something that exists inherently cannot result from causes and conditions. But without dependence on such causes and conditions, causality can never operate."[2] Nothing can be its own cause, and the result of an exterior cause is necessarily impermanent. Buddhism thus concludes that reality is constantly changing, not just in the visible world, but also at the level of the infinitely small, in time as well as space. The idea that quarks are matter's "immutable basic building blocks" is thus simply a mental construct. Quarks are as impermanent as the objects that they allegedly form.

T: I'd even go so far as to say that quarks are inventions we've come up with to explain the behavior of subatomic particles.

M:—And so reality isn't as solid as we think.

T: In order to cover all bases, I should also mention string theory, which we've discussed before. This theory describes quarks not as mathematical points, but as infinitesimally small vibrating strings. It aims to unify general relativity, which describes the infinitely large, and quantum mechanics, which describes the infinitely small. The apparent incompatibility of these two great theories is one of the main stumbling blocks that prevents us from getting a better understanding of the universe. If they could be reconciled, we would then have a theory of quantum gravity, which would let us unite nature's four forces (electromagnetism, the weak and strong nuclear forces, and gravity) into one superforce. String theory, which is also known as the theory of superstrings, has been proposed as this ultimate theory, allowing us to describe all the phenomena in the universe.

According to this theory, particles aren't the fundamental elements, but are vibrations of infinitely small strings of energy, measuring 10^{-33} centimeters, which is none other than the Planck length. The length of one of these pieces of string in comparison to an atom is equivalent to the size of a tree compared to the universe. Particles of matter and light that transmit forces bring together the world's different parts so they interact with one another and change. For instance, the photon transmits the electromagnetic force, while the graviton transmits gravity. All these particles could just be the various manifestations of these strings. Just as the strings of a violin vibrate, these strings vibrate and generate tones and harmonics that are detected by our measuring instruments as protons, neutrons, electrons, and so on.

The energy of the vibration determines the particle's mass. The more energetic the vibration, the greater the mass. In the same way, a particle's electric charge and spin are determined by how the string vibrates. The strings are all basically the same. All that varies is how they vibrate. For example, a proton is simply a trio of vibrating strings, each of which corresponds to a quark. Just as musicians charm us by playing a piece by Brahms, the combined vibrations of these three strings produce the music of a proton. When our measuring apparatus captures it, the music comes out as a mass, a positive electric charge, and a spin. The music of an atom, which is made up of protons, neutrons, and electrons, is played by even more musicians in an even larger orchestra. In this way, strings sing and vibrate all around us, and the universe is in fact a vast symphony.

M: These strings are obviously not inherently protons, neutrons, or electrons. This confirms Buddhism's analysis of matter: an object's characteristics do not belong to it. What exists is a stream of constant transformations that appear in various forms.

T: Particles lose their inherent existence, given that the same strings can appear in different guises when they vibrate at different frequencies. If one of them vibrates in a certain way, it appears as a photon. If it then changes its tune, it becomes a graviton.

M: If particles are just vibrating strings, do the strings have a permanent existence?

T: The strings replace quarks as basic entities, but they can appear either as strings or as waves. Instead of being dimensionless mathematical points, they are shaped like infinitely thin pieces of string apparently existing in only one dimension, rather like tiny pieces of spaghetti. They are so small that even our most accurate instruments see them as points. But they also have "hidden" dimensions. According to one version of the theory, the strings exist in a ten-dimensional universe, with nine space dimensions and one time dimension. In another version, the universe has twenty-six dimensions, twenty-five of them being spatial and one being temporal. We can see only three spatial dimensions; the other six or twenty-two dimensions are shrunk so tightly (to the Planck size of 10^{-33} centimeters) that they can't be seen.

M: Can you explain more about what these vibrating strings are supposed to be like?

T: We can say that a string can be described in terms of its energy (or frequency) and its tension (like the tension of a violin string). Two strings vibrating in the same way produce different particles if their tension is different, because that means they no longer have the same energy.

M: And are these strings indivisible, or can they be cut and spliced together?

T: Strings don't lead a quiet, lonely life. They move, interact, join up, and subdivide, but they can't measure less than 10^{-33} centimeters, which is Planck's length. The ends of a string can be free or connected together into a loop. Or else two different strings can join to form a single loop.

M: Are these strings truly individual entities, then? In Buddhism, the apparent independence of objects necessarily ends up fading away upon closer analysis because of the nonreality of distinct entities. But this string theory is still based on an idea of being able to localize a tangible reality, with an inherent, separate existence.

T: Absolutely. Particles are no longer seen as having intrinsic substance, but that substance has been passed on to strings.

M: Because of our need to reify, aren't we giving a false substance to ideas that are really just mathematical tools here? Not long ago, I had the chance to talk to physicist Brian Greene, a specialist in superstring theory.[3] I asked him if a string could exist independently, without being linked to the rest of the universe. In his words, "Perhaps there really is no notion of a separation between strings and the universe they inhabit. The latter really is a reflection of the former."

T: If string theory is correct, then there is no notion of the universe without strings.

M: We can still wonder why the strings vibrate.

T: According to quantum uncertainty, a string can never be completely still. If it were, then we could know simultaneously its exact position and speed, which is impossible.

M: Okay, so why is there a change in the way a string vibrates? If that is its only characteristic, then it apparently has no immutable properties. To quote Brian Greene again: "Strings can interact with each other, and through these interactions, their vibrational patterns are generally affected." A particle's properties (charge, mass, and spin) depend entirely on the string's vibration, and so aren't inherent to the particle. But, given that the vibration can change, it would also seem that they aren't inherent to the string. When I asked Brian Greene about this, he replied, "Perhaps you are right. The string is like a chameleon. It can 'look' like any particle since all that a particle is (according to string theory) is the pattern of vibration of its internal string."

A reality consisting of self-generating elements, whether they be particles or strings, would imply that there were immutable entities. But, while these strings do apparently simplify things by knocking out any idea of particles having intrinsic properties, they aren't fundamental unchanging entities. They may measure Planck length, but according to Brian Greene, there's no reason why Planck length shouldn't be different in another universe. I think that the Buddhist refutation of a separate, localized, and fragmentary reality can be applied as much to strings as it can to individual, indivisible particles. It emphasizes the vanity of looking for the "final foundation stone" of reality.

T: Of course, all of this about strings may prove to be only a beautiful but flawed idea. For the moment, superstring theory has practically no chance of being checked experimentally. To do so, we'd have to produce far more energy than even today's most powerful particle accelerators can manage. What's more, it's wrapped up in an increasingly dense mathematical fog and is drifting farther and farther away from reality. Physics that hasn't been tied down experimentally is really just metaphysics.

But I don't mean to suggest that I'm not intrigued by the idea that impermanence is at the heart of the nature of reality. In fact, in addition to the impermanence we apparently see at the subatomic level, we see impermanence also at the level of the entire universe. The notion of perpetual change is central to modern cosmology.

The idea of cosmic evolution—that the universe is constantly evolving—is implied by the Big Bang theory, though the idea of evolution only really took over following the discovery of the cosmic background radiation in 1965. In the 1950s, the steady-state theory was favored. It claimed that the universe was immutable and that, on average, it changed neither in time nor in space.[4]

Interestingly, the opposition between the ideas of a changeless world and one that is constantly evolving actually goes back much further, having originated in the fifth century B.C.E. with the Greek philosophers Heraclitus and Parmenides. Heraclitus thought that the universe was in perpetual motion and that everything moved and flowed without beginning or end. Meanwhile, Parmenides declared that movement was incompatible with Being, which was One, continuous and eternal.

M: In Parmenides' opinion, if things changed, then the appearance of something that previously didn't exist would become possible. But something that doesn't exist can't start to exist. So change is impossible. This viewpoint is typical of a philosophy that is rooted in the real existence of things. According to this way of thinking, all results must exist already within their causes, because nothing new can emerge. Buddhism's answer is as follows (in the terms in which it was given to Hindu philosophers with similar ideas to those of Parmenides): "If results exist within their cause, let them rather buy the cotton grains to wear. If the result was present in the cause and indistinguishable from it but not manifest, with the money you spend on cotton cloth, buy some cotton seeds and clothe yourselves with them! They too will serve the purpose of cotton cloth, protecting you from the cold and wind, for, as you maintain, the cloth aspect exists in the seed."

T: Aristotle then incorporated both ideas of change and of stability in his cosmogonic system. According to him, change takes place on Earth and on the Moon, because there imperfection dominates in the form of life, aging, and death. But changelessness can be found on the other planets, the Sun, and the stars, because they are perfect, immutable, and eternal. In the seventeenth century, Newton abolished the Aristotelian distinction between Heaven and Earth with his theory of universal gravitation, according to which the same law of gravity determines both the fall of an apple in an orchard and the orbit of the planets around the Sun. So Heaven is as changing as the Earth.

M: However, in his *Opticks*[5] he defended the idea of permanence at the heart of reality.

T: It's true that the founding fathers of Western science, from Galileo to Newton and Kepler, all accepted the idea that God the Creator was responsible for a perfect, unchanging, and eternal universe. But as science progressed, it revealed that the universe was constantly changing, and the idea that it is immutable is now untenable. Stars are born, live out their lives by burning up their hydrogen and helium, then die and throw out into space their gases enriched with the chemical elements produced by their nuclear alchemy. Then these gases collapse under the force of gravity, thus giving birth to a new generation of stars, and so on. These cycles of life and death last several million or even several billion years. Our Sun, which appeared four and a half billion years ago, or eleven and a half billion years after the Big Bang, is a third-generation star. So the galaxies, which are made up of hundreds of billions of evolving stars, must change too.

What's more, nothing is motionless in space. Gravity sees to it that all the structures in the universe, such as stars and galaxies, attract each other and "fall" toward each other. These in-falling movements must be added to general expansion. In this way, our Earth is taking part in a fantastic cosmic ballet. First, it pulls us through space at a speed of nearly twenty miles per second during its annual journey around the Sun. The Sun then drags the Earth with it during its voyage through the Milky Way at a speed of 140 miles per second. The Milky Way is falling in turn at approximately fifty-five miles per second toward Andromeda. And there's more to come. The Local Group that contains our galaxy and Andromeda is falling at about 375 miles per second toward the Virgo cluster of galaxies, which is in turn moving toward a large complex of galaxies called the Great Attractor. Aristotle's static, immutable Heaven is dead and gone. Everything is changing and nothing is permanent. There can be no doubt that impermanence is all around us.

M: What are the consequences of this understanding for our lives? Buddhism and physics have different aims. Physics stops at the description of phenomena. Buddhism's purpose is to lessen our attachment to the reality we see before us: beings, events, things, even ourselves. For, if we attach ourselves to things as if they were permanent and solid, we think that they inherently have the power to make us happy or miserable. So it is that we give objects characteristics—such as "mine" or "theirs," "beautiful" or "ugly," "pleasant" or "unpleasant"—which are simply conceptual labels. As Dharmakirti explained:

> *When "self" occurs, so too the thought of "other."*
> *From "self-and-other" both, attachment and aversion come.*
> *These two combined*
> *Are source of every ill.*[6]

In this context, Buddhists don't want to determine the mass and the charge of particles, but instead to break down the notion that things are permanent and solid, so as to liberate us from the vicious circle of illusions that cause our suffering.

In Buddhism, the next step is to incorporate this notion of impermanence into our outlook. The Buddha said, "Of all footprints, the elephant's are outstanding; just so, of all subjects of meditation for a follower of the Buddhas, the idea of impermanence is unsurpassed."[7] The impermanence of the macroscopic world is obvious to everyone, but reflection about subtle impermanence has deeper consequences. Phenomena contain the seeds of their own transformation, and the universe can contain no immutable entities. It's this very malleability of phenomena and of consciousness that allows us to undertake the process of transformation that finally leads to enlightenment.

EACH TO HIS OWN REALITY

7

WHEN THE SNOWS OF KNOWLEDGE MELT

Despite the radical discoveries of quantum mechanics concerning the nature of reality, many physicists continue to believe in "material realism," the view that there is a solid reality that can be described in terms of either elementary particles or superstrings. Others, who have grappled more with the philosophical implications of quantum mechanics, argue that the quantum paradoxes indicate that the ultimate nature of reality will forever remain veiled. But is this idea of a veiled reality really so different from the materialist view? Buddhism contends that if we want to grasp the true nature of reality, we must engage much more fully with the philosophical conundrums that quantum physics has revealed.

MATTHIEU: Most people think that in order to explain the coherence of the phenomenal appearance, it is necessary to postulate a wholly independent, extra-mental world, which underpins appearances. To assume a substrate beneath appearance may seem rational, but it mostly rests upon an ingrained and unexamined habit that is not only a fundamental misapprehension of reality, but is also the source of frustration and misery. Even though quantum mechanics refutes the existence of a "true" reality existing on its own, "out there," some scientists continue stubbornly to look for it, no doubt because of their cultural conditioning.

This is very different from Buddhism. Many of the paradoxes of quantum mechanics result from the fact that we are constantly trying to force it into the view of reality established by Western philosophy. As W. H. Zurek wrote, "The only 'failure' of quantum theory is its inability to provide a natural framework for our prejudices."[1] Buddhism's vision of the world is a middle way between solid reality and nothingness. This conception of the world resolves many of the paradoxes of contemporary physics. It thus offers a coherent framework for thought and action in the modern world.

THUAN: Scientists carry out their work in a specific social and cultural context. Even if they are reticent to admit it, they can't help sharing, consciously or unconsciously, the metaphysical prejudices of the society in which they live. It's true that Western science has been dominated by a metaphysical vision of a solid, reified reality. I don't think it's any coincidence that the founding fathers of quantum physics, such as Bohr and Schrödinger, argued for a marriage between Western scientific thinking and Eastern philosophy. In Eastern thought they saw a possible way out from the numerous paradoxes that quantum mechanics produces when understood from a Western point of view. In Heisenberg's opinion, "The great scientific contribution in theoretical physics that has come from Japan since the last war may be an indication for a certain relationship between philosophical ideas in the tradition of the Far East and the philosophical substance of quantum theory. It may be easier to adapt oneself to the quantum-theoretical concept of reality when one has not gone through the naive materialistic way of thinking that still prevailed in Europe in the first decades of this century."[2] And Bohr went on, "For a parallel to the lesson of atomic theory . . . we must in fact turn to quite other branches of science, such as psychology, or even to that kind of epistemological problems with which already thinkers like Buddha and Lao-tse have been confronted, when trying to harmonize our position as spectator and actor in the great drama of existence."[3]

Not many physicists, however, have concerned themselves with the philosophical consequences of quantum mechanics. To most researchers, quantum mechanics is simply a theory that works particularly well. It's an unrivaled way to describe the way matter behaves on a subatomic level, and how it interacts with light. Quantum mechanics is also considered a useful tool that allows us to make transistors, lasers, chips, computers, and other, even more extraordinary instruments that have modified and will continue to radically change our way of life. Most researchers go no further than that, and don't bother about the philosophical implications of their science.

M: I recently jotted down some comments made by the physicist Jean-Marc Levy-Leblond, which confirm what you say:

> *There should be no misunderstanding about the wide-ranging agreement between physicists about most of their theories, from cosmology to particle physics, passing by statistical mechanics. This just concerns the theoretical mechanisms, that is to say the various mathematical forms that are used to describe our experience of the world, and the calculation methods which allow us to produce explanations or predictions based on our observations. . . . But this consensus leaves open a whole series of questions about how to interpret these theories and the meaning of our concepts. . . . Behind the scientific community's unified façade are deep intellectual divergences, which are all the more important given that they are rarely expressed. . . . This range of conceptions is generally masked by the indifference or the caution that most researchers display about anything that lies outside their field of specialized study.*[4]

T: This is definitely not a healthy situation. While quantum mechanics is well advanced when it comes to explaining the behavior of matter, the exposition of its philosophical basis has hardly progressed. That said, a handful of physicists have made philosophical conjectures, for example Bernard d'Espagnat and Michel Bitbol in France. In order to explain certain strange aspects of quantum mechanics, such as "nonseparability" (which came up when we were discussing the EPR experiment), d'Espagnat has introduced the concept of "veiled reality,"[5] which we've discussed. To his mind, science might describe empirical reality extremely well, but it can give us only glimpses of an independent, veiled reality that is not part of normal space-time. This reality can't currently be described by physics and eludes all the conceptual frameworks that we can devise. Nonseparability destroys any attempt to describe that independent reality, and casts serious doubt over the inherent existence of our space-time.

Speaking of materialist realism, Heisenberg wrote, "One was led to the tacit assumption that there existed an objective course of events in space and time, independent of observation; further, that space and time were categories of classification of all events, completely independent of each other, and thus represented an objective reality, which was the same to all men."[6]

M: The idea that there's an inherent substratum that exists beneath the veil of our perceptions has been much discussed by Buddhist philosophers. They have concluded that our concepts can't deal with reality. Since it's neither inherently existent nor nonexistent, the true nature of reality necessarily escapes ordinary intellects. This does not exclude, however, achieving a correct understanding of reality, as the union of appearances and emptiness, in an experiential way that transcends the ordinary conceptual mind.

In any case, one cannot conclude that there is a solid reality simply because we constantly observe its apparent properties.

T: In fact, neurobiology has shown that "reality" only appears the same to members of the same species, equipped with the same neuronal system. Other species have different perceptions of the world. For example, research done on fish, birds, and insects has proved that they see shapes and colors quite differently from us. So reality is unavoidably modified by the neuronal system that perceives it.

M: Experiments have never proved that what we observe exists inherently and has intrinsic characteristics. Experiments are just events. We can see two moons a thousand times by pressing our fingers over our eyeballs, without the second moon being any more real for all our efforts!

T: Bernard d'Espagnat has compared reality to a rainbow seen by the inhabitants of an island in the middle of a river. That multicolored arc looks as real to them as all of the things under it. At one end of the rainbow there's a poplar, and at the other the roof of a farmhouse. The island inhabitants are convinced the rainbow exists. They think that it would still be there, in exactly the same place, even if they closed their eyes or quite simply vanished. And yet, if they could leave their island and drive around in a car while keeping their eyes on the rainbow, they'd see that its position wasn't fixed, and that its two ends weren't always over the poplar and the farmhouse. Its position depends on where the observer is. This comparison shows that even macroscopic objects have no intrinsic existence, and that the observer plays a vital part in how they are perceived.

M: It's true that no theory can ever account for a reality that exists totally apart from the act of perception. Whatever the basis of perception may be, it cannot possibly be the same as what we perceive. Moreover, since it lies beyond the scope of perception (even the extended perception afforded by scientific instruments), its nature must be a matter of conjecture. In fact, to postulate the existence of a *truer,* veiled reality is pure metaphysics. Given that this reality is utterly inaccessible to thought, it would be as unreal as a geometric figure that was simultaneously a circle and a square.

I'm interested in d'Espagnat's rainbow argument, because the rainbow example is often used in Buddhism, though in a somewhat different way. We say that ordinary beings that remained attached to solid existence are like children running after a rainbow in the hope of grabbing it and wrapping themselves up in it. The rainbow, which is luminous but intangible, symbolizes the union between emptiness and phenomena, and also interdependence. It's formed by the conjunction of a curtain of rain and sunbeams, but nothing in fact is created. It disappears as soon as one of these two elements is lacking, but nothing then has really ceased to exist. The conditions for the appearance of a rainbow don't require the backing of a "veiled reality." My teacher, Khyentse Rinpoche, linked this understanding of the nature of phenomena and the contemplative life in the following way:

> *All phenomena of samsara and nirvana arise like a rainbow, and like a rainbow they are devoid of any tangible existence. Once you have recognized the true nature of reality, which is empty and at the same time appears as the phenomenal world, your mind will cease to be under the power of delusion. If you know how to leave your thoughts free to dissolve by themselves as they arise, they will cross your mind as a bird crosses the sky without leaving any trace.*[7]

So, does "reality" lie forever beyond our knowledge? Yes, if we persist in trying to make inherently existing "things" emerge from the world of phenomena. No, if we try to grasp their ultimate nature and lack of inherent existence.

T: But if there is no reality behind the world of phenomena, why do we all perceive more or less the same thing? When I observe the universe, the galaxies and stars through my telescope, I see and measure the same reality, the same properties of light, the same outward motion of the galaxies, the same brilliance and the same colors of the stars as any other observer of the universe. Physical constants change neither in time nor in space. Since telescopes are also time machines that allow us to go back into the past of the universe, I can check that the mass and charge of electrons in the galaxies I'm observing are the same as on our little planet Earth, even though their light began its intergalactic journey ten billion years ago.

If the properties of macroscopic objects—for example, a rainbow in the sky—are partly mental constructs, how do you explain why the various experiments we carry out produce the same results and that we all agree about their nature? How can you explain the subjective agreement between people about what we see in the everyday world?

M: This agreed-upon vision of reality is what Buddhism calls "relative truth accepted by consensus." The world is definitely not random or arbitrary. Even though phenomena have no autonomous reality, they do exist as "simple appearances" that can be called by various names and, since they aren't pure nothingness, can function and interact according to the laws of causality. Those laws explain the numbers, constants, and properties that are revealed by our measurements and calculations. In this context, it's understandable that the mass of the phenomenon which we define as an electron is invariable. So it's only normal that humans measure it in the same way. On the other hand, what is far from obvious is that all the characteristics we attribute to an electron are intrinsic properties that would be perceived in exactly the same way by other beings.

Why do members of the same species see phenomena in practically the same manner? When we say, "Each time I look at this object, I see it in the same way, and so does everyone else," this in no way proves that its observable properties are an integral part of its nature. A perception's apparent stability derives from the continuous interaction between consciousness and a particular set of phenomena. The fact that humans all view the world in practically the same way is explained by the fact that our consciousness and bodies are constructed similarly. But a human's world differs radically from that of an insect, which is different again from that of a bird. Buddhist texts give the example of a glass of water. We perceive it as a drink, or as a means to wash, whereas it would terrify a rabies patient, look like a set of molecules for a scientist using an electronic microscope, or like a dwelling space for a fish. It may appear as fire or anything else to other kinds of beings we can't conceive of.[8]

T: Neurobiologists would certainly agree that different species have different perceptions of the world. For example, the eyes of different species are sensitive to different-colored lights or rays that are invisible to us. A dog can see in the dark because its eyes are more sensitive to infrared light than ours. A pigeon can see ultraviolet rays that we cannot. Bats don't use sight, but perceive objects thanks to the echoes from the high-frequency sounds they produce. Their representation of the world is certainly very different from ours. Biologists think that these differences are due to natural selection. Each species has developed the forms of perception that are best suited to its environment, and best adapted to its survival, reproduction, and proliferation.[9]

M: The relative similarity in how the senses and consciousness for individuals of a species function means that perceptions of the world will be similar, but this does not mean that these perceptions can be considered ultimate. In the *Samadhiraja Sutra,* the Buddha said:

> *Eyes, ears, and nose are not valid cognizers.*
> *Likewise the tongue and the body are not valid cognizers.*
> *If these sense faculties were valid cognizers,*
> *What could the sublime path do for anyone?*[10]

In addition to this, so far as Buddhism is concerned, even consciousness never perceives what we call reality. At the first moment of a perception, our senses capture an object. At the second moment, they create a nonconceptual mental image of a shape, a sound, a taste, a smell, or a touch. When we arrive at the third moment, our mental mechanisms start up, along with our memories and acquired habits, and a multitude of consecutive conscious moments identify the object's image as being this or that. They interpret it and have positive, negative, or neutral feelings about it. Meanwhile, the impermanent object has already changed. Thus normal conceptual consciousness *never* perceives a simultaneous reality. All it can perceive are images of past states.

What's more, a mental image—of a flower, for instance—is deceptive because we don't generally think that it's impermanent and devoid of intrinsic existence. Buddhism calls this "invalid cognition." But it is possible to replace this with a valid cognition, which sees the true nature of the flower (emptiness) and isn't influenced by ordinary concepts. It is said that one of the characteristics of enlightenment is the ability to distinguish between pure, nonconceptual perception and mental images.[11]

T: So, two thousand years before Kant and cognitive science, Buddhism understood that the world we perceive is a mental reconstruction of exterior reality, with the additional notion that this "reality" is never totally distinct from consciousness.

M: The physicist David Bohm summarized this in this way:

> *Reality is what we take to be true. What we take to be true is what we believe. What we believe is based upon our perceptions. What we perceive depends upon what we look for. What we look for depends on what we think. What we think depends on what we perceive. What we perceive determines what we believe. What we believe determines what we take to be true. What we take to be true is our reality.*[12]

No matter how complex our instruments may be, no matter how sophisticated and subtle our theories and calculations, it's still our consciousness that finally interprets our observations. And it does so according to its knowledge and conception of the event under consideration. It's impossible to separate the way consciousness works from the conclusions it makes about an observation. The various aspects that we make out in a phenomenon are determined not only by how we observe, but also by the concepts that we project onto the phenomenon in question.

What's more, a reality that was independent of our senses and concepts would be meaningless to us. What theory could depict a reality that was totally alien to our intellects? How could the characteristics of this reality appear to us, without first having been influenced by the very act of looking for them? This idea also occurred to Henri Poincaré, who wrote, "It is impossible that there is a reality totally independent of the mind that conceives it, sees it, or senses it. Even if it did exist, such a world would be utterly inaccessible to us."[13]

Alan Wallace has summed this problem up neatly:

> *To adopt scientific realism consciously, we must accept a number of underlying premises: (1) there is a physical world that exists*

independently of human experience, (2) it can be grasped by human concepts (mathematical or otherwise), (3) among a potentially infinite number of conceptual systems that can account for observed phenomena, only one is true of reality, (4) science is now approaching that one true theory, and (5) scientists will know when they have found it.[14]

Descriptions produced by the natural sciences bring together observations that have been made, organize them, then predict how they will develop. But they don't point to an autonomous reality.

T: But my question still remains: What is the basis in the "real" world for these varied perceptions?

M: The inseparability of the relative truth of phenomena and the ultimate truth of emptiness is the only possible answer. This is the wisdom perceived by anybody possessing the perfect knowledge of a Buddha.

Neither inherently existing objects nor emptiness alone can provide a basis for the emergence of phenomena. The only basis that stands up to analysis is a "simple appearance," which isn't confined to an exterior object or to interior consciousness.[15] We thus come back to the indivisible union between appearances and emptiness, which transcends the conceptual extremes of existence and nonexistence.

It's obvious that the exterior object we perceive at a given moment isn't a pure invention of our minds. However, the entirety of our "landscape," or the way we perceive the world, merely results from the way our minds have developed and the experiences we have accumulated. This is why members of the same species perceive a more or less similar set of phenomena. The different species experience different parallel "unrealities," which lead to various perceptions of what we call "the same glass of water." But, careful. When we say "the same," this doesn't mean that there's a "real" glass of water positioned behind the semitransparent screen of our senses. Here, "the same" means that the working of various conscious minds during numerous previous existences has led to similar crystallizations. They reflect a process during which concepts formed by our perceptions become fixed, and appear more or less similar depending on how those different conscious minds have functioned during all that time. This of course runs against the commonsense view, which insists that there is a world that is totally exterior to us.

T: Surely you don't mean that if we weren't here, then the world wouldn't exist!

M: No, of course not! Although there is a school of Buddhism called "Mind Alone" *(Cittamatra),* which claims that only the mind has an ultimate reality and phenomena are mere mental projections, its position was refuted by the Middle Way *(Madhyamaka),* which is considered to be the deepest form of philosophy. According to the Madhyamaka school, the right way to look at reality is in terms of the interdependence between conscious and unconscious phenomena, neither of which exists in absolute terms. It's obvious that the world doesn't vanish when we go to sleep or pass out. Nevertheless, the interaction between our consciousness and "exterior" phenomena forms a network of special relationships that define "our" world. And this world does fade away when one of its elements—consciousness, for example—is missing.

A tree branch breaks even if we aren't there to see it. But the world of phenomena has to be viewed in a larger perspective. It has come about through a long fermentation caused by the coexistence of the consciousness of beings and the infinite potential that phenomena owe to their emptiness. My teacher, Khyentse Rinpoche, put it this way: "When a reflection appears in a mirror, you cannot say that it is a part of the mirror, nor that it lies elsewhere. In the same way, perceptions of exterior phenomena take place neither in the mind nor outside. Phenomena are not really existent or nonexistent. So the realization of the ultimate nature of things lies beyond the concepts of being or nonbeing."

This is why Nagarjuna concluded, "If I assert anything, then I am at fault. But since I assert nothing, I alone am faultless!"[16] To assert nothing isn't a denial of reason. It's the realization that the ultimate nature of phenomena can't be established by concepts, enclosed in definitions or wrapped up in the categories of solid reality or nothingness.

We've come back to the notions of relative and absolute truth here. Relative truth is the way phenomena appear to us, with identifiable characteristics. Absolute truth shows that these characteristics have no inherent existence, which implies that the ultimate nature of phenomena, which is emptiness, lies above and beyond any description or concept. It is said in the *Prajnaparamita:* "To understand perfectly that things have no reality in themselves is the practice of supreme, transcendent knowledge."

T: Scientific knowledge is more like the relative truth you describe. What makes experimental science possible is the agreement between people and the similarity of perceptions of the world of phenomena. Observations and experiments must be repeatable and confirmed by different research teams using other instruments and techniques. Only then are results considered to be valid. Thus the aim of physics is not to describe an inherent reality but "communicable human experience," that is to say, observations and measurements.

M: This is a very important point, because it breaks through the materialistic view that there is an essential difference between objective and subjective reality. This also allows us to put science in its proper place. It's quite simply an evaluation and organization of the *relationships* which make up phenomena on the conventional level, and the ability to have an increasing influence on these phenomena. Buddhism calls such phenomena "events." The literal meaning of *samskara,* the Sanskrit word for "things" or "aggregates," is "event" or "action."

T: This etymology is akin to what I said earlier about Bohr's interpretation of quantum mechanics: the notion of "object" is subordinate to the "measurement," hence to an event.

M: This is also the view of Buddhism, which states that reality is always determined by the interaction between the observer and the observed. Complementarity between the whole and the parts means that it is sometimes the "part" aspect that is revealed, and sometimes the global aspect. All the observer does is to isolate a certain spectrum of aspects, which have no more reality than a particular interaction between the observation and global nature, that is to say between a consciousness and the whole of which it is a part. What we call reality is thus a "viewpoint" of consciousness.

For an ordinary mind, there's a difference between the way things seem and their true nature. In terms of our personal experience, this leads to a mixture of suffering and happiness. At the end of our journey of discovery, we directly perceive the ultimate nature of phenomena. This leads to immutable wisdom in which all disparity between appearances and reality has vanished. From this wisdom, a limitless compassion spontaneously arises for all of those beings whose ignorance plunges them into endless suffering.

QUESTIONS OF TIME 8

Modern physics has progressed from Newton's concept of absolute, universal time to Einstein's relative, malleable time, which can slow down or speed up according to an observer's motion or the strength of gravity. Time has lost its universality: one person's past can be another's future. What's the difference between physical time and the psychological time that we experience? Is "time's arrow," the sense that time moves in one, forward direction, real or is the direction of time a mental construct?

THUAN: Gaston Bachelard said that "meditating on time is the first step toward metaphysics." And time is certainly not an easy notion to grasp. Saint Augustine observed in the fourth century: "What is time? If no one asks me, then I know. But if someone questions me about it and I try to explain it, then I no longer know."[1]

Time plays a vital role not only in metaphysics, but also in physics. When studying nature, physicists are constantly being confronted by questions about time. At first this may sound paradoxical, since time measures the ephemeral and physicists are searching for laws, that is to say invariable and immutable relationships. And yet the notion of time keeps cropping up in physics.

In the sixteenth century, Galileo introduced the idea of time as a basic physical dimension in his studies of moving objects. But it was Newton in the seventeenth century, with his laws of mechanics, who provided the first explicit definition of time. He defined the movement of bodies in space by specifying their positions and speeds at successive moments in time. Newtonian time was absolute and universal; it flowed in the same way for everyone, and each observer in the universe shared the same past, present, and future. Space and time were strictly distinct. Time passed by without interacting in any way with space.

In 1905 the idea of absolute time was questioned by Einstein in his article on the Special Theory of Relativity. According to Einstein, time was no longer independent of the universe in which it was supposedly flowing invariably but became elastic and dependent on the motion of an observer. The faster we go, the slower time elapses. For example, for someone moving in a spaceship at 87 percent of the speed of light, time would slow down by one-half. He would age half as quickly as his twin on Earth. This age difference would be perfectly real. His twin would have more wrinkles and more white hair. His heart would have beaten more often and he would have eaten more meals, drunk more wine, and read more books. This example is known as Langevin's twin paradox (named after the French physicist who invented it). But it's only a paradox according to our sometimes misleading common sense. The Theory of Relativity precisely accounts for this slowing down of time. This is imperceptible when it comes to speeds we encounter in our daily life. But it becomes important with speeds that are near the speed of light (186,500 miles per second). At 99 percent of the speed of light, time slows down seven times. At 99.9 percent, 22.4 times. This slowing down of time is no mental game; it has been observed for particles launched at high speeds in accelerators. They live longer (before disintegrating) than when they're still, and always according to the proportion predicted by Einstein.

Einstein also revealed that time and space do not lead separate lives. He turned time and space into a tightly knit couple. Space is also elastic. The behavior of both partners is always complementary. When time stretches out and passes more slowly, space contracts. If one of our twins speeds along aboard a spaceship traveling at 87 percent of the speed of light, not only does he age twice as slowly, but his space contracts: as far as his twin on Earth is concerned, the spaceship looks as if it's shrunk by one-half. These deformations of time and space can be seen as the transmutation of space into time, and vice versa. Shrinking space changes into lengthening time.

Time is slowed down not only by speed, but also by gravity. This is the radical notion that Einstein announced in his 1915 General Theory of Relativity. When next to a black hole, which has a colossal gravity, an astronaut's watch would slow down in comparison to one on Earth. The slowing down of time by gravity has also been verified experimentally. Physicists have succeeded in measuring a fractional change of time of 2.5 millionths of a billionth of a second between the top and bottom of a seventy-five-foot tower on the Harvard campus. The clock at the bottom of the tower goes slower because it is closer to the center of the Earth and gravity is slightly larger there, while its counterpart at the top of the tower goes faster because it is slightly farther away from Earth's center and gravity is slightly smaller there. The slowing down of the clock at the bottom as compared to the clock at the top corresponds to a time lag of one second in 100 million years, and is exactly the value predicted by Einstein's General Relativity theory.

The elastic nature of time has a fundamental consequence. If time loses its universality, then it isn't the same for everyone. My present could be someone else's past and a third person's future, if the other two are moving in relationship to me. Since the concept of simultaneity loses its meaning, the word "now" becomes ambiguous. And if, for someone else, my future already exists and my past is still present, then all moments are equally valid; there's no longer a privileged instant that we can call the "present." Einstein, whom the British-Austrian philosopher Karl Popper called "the new Parmenides," thought that the passing of time was an illusion. As if to allay his grief, he expressed this viewpoint in a condolence letter written in 1955 after the death of his lifelong friend Michele Besso and less than a month before his own death: "Now he has departed from this strange world a little ahead of me. That signifies nothing. For us believing physicists the distinction between past, present, and future is only a stubbornly persistent illusion."[2] So, for the modern physicist, time no longer flows. It's quite simply there, motionless, like a straight line extending to infinity in both directions.

MATTHIEU: The classic view in Buddhism is that *physical* and *absolute* time is a mere concept with no inherent existence. Time belongs to the relative truth of the world of phenomena, of experience. Time has no inherent existence because it does not exist in the present moment. It's impossible to pin it down at the beginning, during, or at the end of a given period. If we divide a period of time into a beginning, a middle, and an end, then it's clear that the *whole* doesn't exist in any one of these three parts. Nor does the period exist apart from its beginning, middle, and end. Thus "a period of time" is a purely conventional notion. Time, like space, exists only in relation to our experience; it is a concept linked to a perceptible change.

T: We must make a distinction between subjective (or psychological) time and physical time, which is supposedly objective and which flows uniformly and is independent from our consciousness. Physical time is clock time. It's measured in terms of a regular motion, such as the vibration of an atom[3] or the Earth's rotation. That's why there's no sense talking about time (or space) before the birth of the universe. There's no motion to be measured. In the Big Bang theory, time and space were born simultaneously with the universe. Saint Augustine, too, considered that time started with the world. He thought it ridiculous that God should have waited an infinite length of time before creating the world. In his opinion, the world and time arrived together. The world wasn't created *in* time, but *along with* time. This view of time anticipates modern cosmology in a quite remarkable way.

But physical time is different from subjective time, which is the time we experience in our lives, time that doesn't flow uniformly. We all experience its elasticity. The same play can last an eternity for someone who's bored to tears, and pass in a twinkling of an eye for his neighbor. A minute of boredom or fear can seem like a century, while a moment of happiness shoots by. What's more, we all notice that the older we get, the faster time seems to pass. This acceleration of time with age has been demonstrated in studies on the growth of plants and animals. The greater the age, the shorter the "physiological" duration.

This dissimilarity between lived time and physical time has been a constant theme in the history of thought. For the pre-Socratic philosophers, time was identified with a movement, like physical time. Heraclitus said that "time is a child playing backgammon" (time is defined by the movement of the pieces on the board). For Aristotle, time was "the number of movement," but he was already wondering: "It is difficult to know if, in the soul, time exists or not." In the fourth century, Saint Augustine rejected Aristotle's ideas: "Time is not the movement of a body." He affirmed the existential (or psychological) aspect of time: time flows only in the soul, given that the object of expectation (the future) becomes the object of attention (the present) before becoming the object of memory (the past). This was the position also adopted by Edmund Husserl, the twentieth-century German philosopher.

M: Kant also said that the concepts of time and space are a matter of our relationship with nature, and are not integral parts of nature itself: "Time is merely a subjective condition of our intuition, it is nothing outside its subject." An equivalent idea has been suggested by Buddhist philosophers, who state that time has no ultimate reality and has no existence outside of phenomena and their observers. Seeing an "arrow" to the direction of time merely reflects our attachment to relative truth.

By speaking of the "beginning" of time, we can easily see how illusory it is. For, once we've established the idea of a "beginning," we quite naturally ask what happened "before" this beginning. This meaningless question clearly shows that we're not dealing with the beginning of a reality, but of a mental construct. Instead of saying, like Einstein, that time is *always* there, like a motionless dimension, we would say that it's *never* there, which almost comes down to the same thing in terms of the illusory passing of time.

T: Saint Augustine also said that the only time was lived time. Outside the psychological states of memory, attention, and expectation, time is nothing. So, given that everything exists in time, nothing exists. In other words, everything is merely an appearance.

Psychological time nevertheless seems very real. For us, "time passes" or it "flows" like a river. From our motionless boat, anchored in the present, we watch the river of time flowing back into the past and arriving from the future. We give time a spatial dimension, and it's this picture of time moving in space that gives us the sensation of the past, present, and future. The past is full of memories and the future of expectations.

We probably feel the passing of time because of our cerebral activity. Data concerning the external world are transmitted by our sensory organs to our brain, which incorporates them into a mental picture. This cerebral activity brings into play simultaneously several separate regions of the brain with different functions. According to the neurobiologist Francisco Varela, it's the complexity of bringing together and integrating these various parts of the brain that gives us the sensation of time. This orchestrated, synchronous activity of large, discrete sets of neurons, among the hundreds of billions in the human brain, creates what scientists call an "emergent" biological state, that is to say a state, in this case the consciousness of time, that is more than the sum of its parts. Since this state lasts from a few tens to a few hundreds of a millisecond, we have the sensation of "now," of a present with a duration. But this synchronization of neurons is unstable and doesn't last. Its instability sets off other synchronous groupings of neurons, producing a succession of emergent states. They then give us the sensation of time passing. Each emergent state forks off from the preceding one, so that the previous one is still present in the succeeding one. This gives us the impression that time is continuous.

One of the most important qualities of psychological time is that it always flows in the same direction and takes us from birth to death. Like an arrow flying ahead after leaving the bow, psychological time never goes backwards. It is this irreversibility that makes us so fear death. We all know that we progress from the cradle to the grave.

M: Buddhism finds substantial value in the phenomenon of psychological time. It helps us to overcome the fear of death and encourages diligence in the work we do to accomplish spiritual change. A practicing Buddhist will not live in fear of death because, by constantly meditating on it, he has prepared himself to accept it with serenity when the moment comes. Gampopa, an eleventh-century Tibetan sage, said, "At first you should be driven by a fear of birth and death like a stag escaping from a trap. In the middle, you should have nothing to regret even if you die, like a farmer who has carefully worked his fields. In the end, you should feel relieved and happy, like a person who has just completed a formidable task."[4] A hermit turns over his cup (people do this in Tibet when someone dies) every night in case he doesn't wake up the next morning. He thinks that each moment brings him closer to death. Every time he breathes out, he feels happy to breathe in once more. In his "Letter to a Friend" *(Srulekha),* Nagarjuna says:

> *If this life assailed by many ills*
> *Is yet more fragile than a bubble on the stream,*
> *How wonderful it is to wake from sleep*
> *And having loosed one's breath, to breathe in once again!*

In the "Chapters Spoke with Intention" *(Udanavarga),* the Buddha declared:

> *All that has been gathered will disperse,*
> *All that is constructed will decline and fall,*
> *All that meets will one day separate,*
> *And all that lives will vanish into death.*

Thus the realization that time passes swiftly and irreversibly acts like a spur to our diligence. Padmasambhava, the master who introduced Buddhism in Tibet, declared:

> *Like streams and torrents flowing to the sea,*
> *Like Sun and Moon that seek the western hills,*
> *Like days and nights, the hours and instants flying,*
> *This life of ours goes by relentlessly.*

The Buddha used the image of an athlete catching four arrows shot at the same time by four archers facing in different directions. "And yet," he went on, "even faster is the passing of time and the approach of death." For a practicing Buddhist, then, time is his most precious commodity. Not a single moment should be wasted in the indifference of someone who has forgotten that he will die.

But this doesn't mean that Buddhism believes time's arrow is real. The past and future have no reality and the present is ungraspable.

T: Indeed, this psychological perception of time passing by our motionless consciousness doesn't agree with the vocabulary of modern physics. For example, if "time flows," a physicist can ask, what would be its speed of flow? An evidently absurd question. I should point out that on the subatomic level, time isn't one-way anymore. In the world of particles, time's arrow disappears and time can flow in either direction. Two converging electrons collide and separate again. If we invert the sequence of events, we still have two electrons colliding and separating again. Both sequences are identical. The physical laws that describe such events do not contain a particular direction of time. A film of the particle world could thus be projected either way around.[5]

M: If, on the level of particles, time has no absolute meaning, how can it start "existing" on the macroscopic level anywhere else than in our minds?

T: As I've already said, Einstein thought that the forward flow of time was a mere illusion. By demolishing the idea of universal time, he abolished the distinction between the past, present, and future. He hoped also to destroy the idea of irreversibility in physics. But time's arrow continues to crop up in other contexts, even in the subatomic world of quantum mechanics,[6] and to dominate the macroscopic world. Just as there is a psychological arrow that always goes forward, so there is a thermodynamic arrow that also goes only in one direction. This is because of the second law of thermodynamics, the science of heat, which states that systems tend toward greater disorder. Entropy, which is a measure of the universe's disorder, can never decrease. We can see examples of increasing disorder when ice melts, or in the stones of a ruined castle. In both cases the initial state was more highly organized than the final state. The ice cube, with its crystal structure, is more ordered than the puddle of water after it's melted. The organization of the castle in its glory days was far greater than the heap of stones that it's now become. In the same way that a change from the past to the future defines the direction of psychological time, so a change from order to disorder defines the direction of thermodynamic time. When Buddhism speaks of impermanence, does it consider that such change has a direction?

M: Buddhism is of course conscious of conventional time's direction. The subtle impermanence of phenomena is quite similar to entropy. The texts say that if a house ages and finally falls into ruins, it is because no phenomena, even the tiniest particle, remain the same. They all carry the seeds of their destruction.

T: The thermodynamic arrow provoked a cry of despair from German physicist Hermann von Helmholtz in 1854: "The universe is dying!" He thought that the increase in entropy that accompanies any natural process would inevitably lead to the end of all creative activity in the universe. Cosmic construction (of planets, stars, galaxies and so on), the works of human genius (Mozart's operas, Monet's *Water Lilies*) would all be buried beneath the remnants of an irrevocably ruined universe.

If the second law of thermodynamics does lead inexorably toward the death and decay of the universe, then why aren't we living in a totally chaotic world? How to explain the organization and harmony of the cosmos? How did the universe ascend the pyramid of complexity? Starting with an energy-filled vacuum, how did it produce elementary particles, galaxies, stars, and planets, and then life and consciousness? Can it be that the second law of thermodynamics breaks down in some parts of the universe? The answer to the last question is no. Thermodynamics doesn't forbid areas of order to emerge in the universe, so long as this creation of local order is compensated for by increased disorder in another location. Let's go back to the example of the ruined castle. A team of builders could reconstruct it. But, to do so, they'd have to eat and thus convert the orderly energy contained in food into the disorderly energy that is dissipated as heat in their bodies. In the end, the builders would create more disorder than the order resulting from the restoration of the castle. The second law of thermodynamics has thus been respected.

The thermodynamic arrow is linked to cosmological time's arrow, which results from the expansion of the universe. As time passes and the galaxies draw farther apart, the universe cools down, spreads out, and becomes less dense. When it was three minutes old, its temperature was several million degrees. After 15 billion years of evolution, this has fallen to the icy level of -270° Celsius, which is the temperature of the cosmic background radiation. Within this bitter cold, the stars are sources of heat and energy thanks to the nuclear reactions taking place in their centers, which are heated to several tens of millions of degrees. The drop in disorder caused by the arrival of complex structures such as galaxies, stars, and planets is compensated for by the disorder resulting from the energy that stars emit into space. So the second law of thermodynamics has been obeyed once again. The thermodynamic and cosmological arrows are thus linked intimately.

But questions concerning the direction of time are far from having been answered, and they remain wrapped up in mystery. If one day the universe reaches its maximum extent and collapses in on itself, will the direction of thermodynamic time, which is connected to universal expansion, flow the other way around in a contracting universe? Will a heap of stones spontaneously organize itself into a beautiful castle? Will psychological time also go in the opposite direction? In fact, if the answer to the last question is yes, the inhabitants of a contracting universe would think that they were in an expanding universe, because their mental processes would also be reversed. In that case, the question of the inversion of time wouldn't really apply to us, except as a mind game.

But to get back to our discussion of Einstein's concept of physical time, that concept clearly faces some problems, and it has had many critics. The French philosopher Henri Bergson couldn't accept Einstein's theory that time is an illusion, without any reality or duration. He thought that time must have a "density." It is this density alone that is compatible with our inner life. Only its duration allows for freedom, creation, progress, novelty, invention, and the workings of the mind. Husserl also spoke of an "incompressible time," which has been confirmed by modern neurobiology. The latter says that time cannot be compressed to zero duration because it takes some finite duration for our neurophysiological processes to operate and give us the sensation of time. Isn't Einstein's physical time overly deterministic and dehumanizing? If everything that will happen is already preordained, what happens to free will and hope?

M: The idea that "everything that will happen is already preordained" is illogical. As we've already seen when discussing a possible Creator, if everything was preordained, then all of the future's causes and conditions should already be present. If that were so, nothing could stop them from expressing themselves at once. If they aren't all present, then there's something still left to add.

T: How can we reconcile the two sorts of time, the physical and the subjective, with one another?

M: Isn't the idea of physical time simply an abstraction of psychological time, born of the distinction we make between what has been done and what is still to be done? Aren't physicists reifying our psychological experience by introducing the notion of physical time?

We can debunk the idea of *physical* time by considering that because an instant has no duration, then several instants have no duration, either. Thus time is merely a label we place on our perceptions of change. If physical time existed in an absolute manner, there would have to be a continuity. This implies a point of contact between the past and the present, and between the present and the future.

T: That's logical.

M: The instant would then be the durationless point where the present met the future. But how, then, could the moment that has just passed by and the present moment have anything in common? If this were so, either the present moment would become the past, or the past moment would become the present. In the same way, the present would have a point of contact with the future, and so the present moment could become the future, or the future moment the present. We would then have an infinite number of past and future moments that could mingle with the present.

T: There's a similar line of reasoning in Aristotle's *Physics:* "If the before and the after were both in one single Instant, how would things be if what was ten thousand years ago were simultaneous with what is happening today?"[7]

But I must remind you that, for neurobiologists such as Varela, the instant, or "now," does in fact have a duration. This can't be less than a few tenths of a millisecond, which is the smallest possible time the neurons need to do their work.

M: The reasoning I just referred to doesn't concern physiological, subjective time, but rather the notion of physical time. My concern is to break down our attachment to a time seen as a "reality in movement." If the passing of time can't be seized in the present instant, which is beyond movement, how do we go from the present to the future? In his *Abhidharmakosha,* Vasubandhu wrote, "Because of the immediate destruction of the instant, there is no real movement, but the production of instants is unbroken."

T: There's an echo of that idea in what Boethius said: "The now that passes produces time, the now that remains produces eternity."[8] For Kant, time can't be disassociated from the thought that perceives it. According to him, time allowed for a succession of events, while space made simultaneity possible. But, like Newton, he considered time and space to be distinct. This is incompatible with the interrelation of time and space that Einstein discovered. Doesn't Buddhism also think that time is distinct from space?

M: Not really. It also refutes the idea that space is a real entity. If we take any region of space, regardless of its extent, can it be anything other than a concept? Space can't be reduced to one of its parts, or be viewed independently from its parts. If the entity "space" corresponds to the entirety of its parts (and if it has an extent, it must have regions or parts), in order to go inside this entity, we'd have to penetrate all of its parts at once, which is impossible. Thus the entity "space" is another mental label with no inherent existence.

T: Is the Buddhist notion of time in any way similar to the Einsteinian conception of a time that doesn't flow? In the Theory of Relativity, physical time is quite simply there, motionless and static. All of space-time is there, containing all the events from the birth of the universe to its death.

M: No, time isn't quite simply there, motionless and static, because it has no reality! Einstein's space-time can't be seen as being absolute. It's just another convention. Time's absolute nature is its emptiness, its lack of inherent existence. This is called "the fourth aspect of time, which transcends the other three" (the past, present, and future). This fourth time is occasionally compared to the present, which, by its very nature, lies outside any notion of duration.

From the contemplative point of view, remaining within "the freshness of the present instant" helps us to recognize the mind's empty and luminous nature and the transparency of the world of phenomena. This nature is immutable, not in the sense of being a sort of permanent entity, but because it is the mind's and phenomena's true mode of existence, beyond any concept of coming and going, being or not being, one or many, beginning or end.

CHAOS AND HARMONY

9

FROM CAUSE TO EFFECT

The findings of relativity theory and quantum physics, as well as those of the relatively new sciences of chaos and complexity theory, have seriously challenged the laws of cause and effect inherited from Newton. Relativity theory tells us that the motion of an observer can modify the temporal succession of events. Quantum mechanics reveals that at the heart of matter, uncertainty rules. Chaos and complexity theory show that the relationship between cause and effect is often anything but linear. Do all of these findings suggest that we must adopt a radically revised notion of cause and effect? And if so, how might this new conception conform to Buddhist ideas of mutual causality, based on interdependence and global reality?

THUAN: One of the most radical findings of Einstein's Theory of Relativity is the elasticity of time—the fact that events can happen at different times according to the movement of an observer of those events. The classic example used to explain this is one of Einstein's thought experiments, of three people in different states of motion, all observing lightning hitting a moving train.

Imagine a train going through a station at high speed. Lightning strikes both ends of the train. Three people are positioned at the level of the middle of the train, but in different places. *A* is on the platform, *B* is on the train, and *C* is on a second train going in the opposite direction. These three people will see the two strikes of lightning in different ways. *A*, standing still on the platform, sees the lightning strike the front and the rear of the train at the same moment of time. *B*, sitting on the train, in the middle of the train, sees the lightning strike the front of the train first and then the rear a fraction of a second later. There's a simple reason for this difference. Because *B* is moving in the direction of the front of the train he is moving toward the lightning bolt that hit the front, and therefore its light has less distance to travel in order to reach *B* than the light from the rear, which has to catch up to *B*. The opposite applies to *C*, who's sitting in the train going in the opposite direction. He sees the lightning strike the rear of the train first, then the front. Who's right? Everyone is, because all three viewpoints are valid. Thus the motion of an observer can modify the temporal succession of events. These differences, which are tiny in the case of our train, would be highly significant for a spaceship traveling at nearly the speed of light.

Given that a sequence of events can be rearranged according to the motion of the observer, a troubling question arises: does Einstein's Special Theory of Relativity cast doubt on the fundamental principle of causality, the rule that a cause must precede an event? Can a result come before a cause? Can an omelette exist before we break the eggs that make it? Can I be born before my mother?

MATTHIEU: But if we accept Einstein's view that the past, present, and future form a continuum—that they are really just different spots on a stretch of time that does not really have a forward direction—then going back in time wouldn't necessarily allow for the alteration of past events. Time reversal would be like rewinding a video, but not rerecording it.

T: To be more precise about Einstein's view, he would have said that each person records a film containing identical scenes that, according to his or her motion, can be placed in different orders. But to address your point, the actual sequence of events can be changed only under highly unusual circumstances. To understand those circumstances, we have to understand the rules about how events are linked causally. For two events to be causally linked, information must be passed from one to the other. This information can be, for example, the position or the state of an object. The fastest speed at which information can travel in the universe is the speed of light. Therefore, in order for two events to be causally linked, light must be able to travel from one to the other in the time that separates those two events. Or we could say this the other way around: For two events not to be causally linked, they must be far enough apart in space or near enough in time so that light can't travel between them in the interval of time that separates them. In both cases, they can't cause each other, so the law of causality does not apply. But if light can travel between them in the interval of time that separates them, then they must follow the principle of causality.

M: Buddhist logicians also say that two simultaneous "real" entities cannot have any causal relation.

T: We can return to the example of the lightning hitting the train in order to understand this better. The order of events in this case is changeable (*B* sees the order in the reverse that *C* does) because the interval between the two events is zero. This is so because *A* sees the lightning bolts strike the front and the back of the train at exactly the same time. Thus light doesn't have time to travel from one bolt to the other. In that case, the two lightning bolts are not causally related, and the order in which they occur can change, depending on the movement of the observer.

On the other hand, if event *b* is preceded by event *a* in a long enough interval so that light has time to travel from *a* to *b*, then *a* always comes before *b* for *all* observers. Light has plenty of time to travel from eggs to the omelette during the interval of time between cracking the eggs and the appearance of the omelette—and for this reason no one will ever see an omelette made before the eggs with which it is made are broken. For the same reason, a child cannot be born before his mother.

Werner Heisenberg explained this finding this way:

As a consequence of the theory of special relativity, two events at distant points cannot have any immediate causal connection if they take place at such times that a light signal starting at the instant of the event on one point reaches the other point only after the time the other event has happened there, and vice versa. In this case the two events may be called simultaneous. Since no action of any kind can reach from the one event at the one point in time to the other event at the other point, the two events are not connected by any causal action. For this reason any action at a distance of the type, say, of the gravitational forces in Newtonian mechanics was not compatible with the theory of special relativity. . . . Therefore, the structure of space and time expressed in the theory of special relativity implied an infinitely sharp boundary between the region of simultaneousness, in which no action could be transmitted, and the other regions, in which a direct action from event to event could take place.[1]

Thus the Special Theory of Relativity is perfectly consistent with the principle of causality. This is just as well, because if the principle of causality wasn't respected, then we'd find ourselves in totally illogical situations. In principle, I could travel back in time and stop my parents from meeting, thus making my birth impossible, which is absurd. This well-known example is what's called the mother (or father) paradox.

M: Don't physicists argue that causality could be reversed if information could travel faster than the speed of light?

T: Contrary to what most people think, relativity doesn't outlaw the existence of particles or phenomena that travel faster than light. In fact, physicists have a name for particles that travel faster than the speed of light, even though they have never been observed. These hypothetical particles are called tachyons, from the Greek *takhos,* meaning "fast." If they really did exist, then they would cause all sorts of physical paradoxes! Traveling faster than light would allow us to go back in time, and would raise the possibility of interfering with causality, such as in the mother paradox.

M: But tachyons could exist only in a theoretical world in which the cause-and-effect relationship was the opposite of ours.

T: That's right. In a world of tachyons, logic as we know it would become meaningless. Effects would come before causes, a nail would be driven home before being hit by a hammer. Einstein was well aware of the consequences of tachyons, and categorically declared in his 1905 paper (in which he presented the Special Theory of Relativity) that traveling faster than light was not permitted. But there's no mathematical clause in the paper that forbids their existence.

What Einstein's theory does explicitly forbid is for any object, or observer, to cross the barrier of the speed of light—in other words, to accelerate from a speed slower than the speed of light to one faster than it. No object, or person, can go from our universe, where all objects move slower than light, to a tachyon universe, where all objects move faster than light.

If an object (or piece of information) could accelerate from a speed slower than light to a speed faster than light, it could catch up to a light beam moving in front of it, and overtake it. The apparent speed of light for an observer on such an object would initially decrease until it became zero, and then increase in the opposite direction, which contradicts the fact that an observer always measures the same speed of light (186,500 miles per second), regardless of his motion. The invariability of the speed of light is in fact one of the basic postulates of special relativity. In the same way, no object can cross the wall the other way, from a speed faster than the speed of light to a speed slower than the speed of light.[2]

M: By stating that causality depends on the speed of light, you seem to be limiting causality to the world of forms, of particles and photons. You also seem to be envisaging a purely linear type of causality in which "*a* leads to *b,* and *b* leads to *c,*" without taking into account the fact that all of the universe's phenomena are inextricably linked together, as shown in quantum mechanics. I don't see why the speed of light defines the range of causality. Heisenberg thought that the principle of causality dictated by the Special Theory of Relativity made a bad fit with the vision of the global nature of phenomena—of global interdependence—revealed by quantum mechanics. Both the experiment of Foucault's pendulum and the EPR experiment have clearly shown that two phenomena can be instantly correlated—that two phenomena can act on one another—without any information having been relayed between them.

T: The fact that there's a zone of causality determined by the speed of light isn't incompatible with the global world revealed by the EPR effect and Foucault's pendulum. As you say, in neither the EPR experiment nor the experiment of Foucault's pendulum is there any transmission of information. In fact, that's the crux of both experiments, to show that it isn't necessary for information to be transferred from one particle to the other, or from the distant universe to the pendulum, in order for it to adjust itself according to the most distant clusters of galaxies. So these experiments do not really contradict the law of causality governing cases in which information is being transferred.

M: And yet it seems that the phenomenon of interdependence as revealed by quantum mechanics overrules the view of causality put forward by physics. According to interdependence, as the Buddha said, "This arises [the observation and behavior of the north particle] because that is [the observation of the other particle at the south that behaves in the same way]." This sort of relationship doesn't imply any transmission of information. The coexistence of phenomena and their unbreakable interdependence thus clearly stand at the heart of causality.

T: Yes, there is a single, global reality. And that's just how the EPR experiment has been interpreted. Quantum mechanics' answer is unambiguous: when two quantum systems interact and split up again, they can no longer be described by two independent wave functions, but only by a global one.

The observer also plays a central role in quantum mechanics. Before measurements are made, we have a wave that is present everywhere, but as soon as we measure it, a particle appears in a particular place. This is called the "collapse" of the wave function.[3]

M: Any measurement, whether it be automatic or voluntary, requires the preparation of equipment. Can we thus conclude that the collapse of the wave function necessitates conscious intervention?

T: In the universe, there are of course plenty of completely unconscious interactions. For example, the fusion of protons in the sun that creates its energy, or a magnet attracting a nail with its electromagnetic force.

What is, then, the difference between measuring apparatus and any other macroscopic object? Why does the former lead to a collapse of the wave function and a choice among the various possibilities? Those who have tried to answer this question can be divided into two main factions.

On the one hand are the "idealists," or "subjectivists," and in particular the Hungarian-American physicist Eugene Wigner. They think that it's the presence of a conscious mind that makes the wave function collapse. Wigner wrote, "Atomic phenomena cannot be described without invoking consciousness. It is the registration of an impression in our consciousness which changes the wave function." But giving the consciousness this major role creates problems. A certain amount of time passes between the moment when the apparatus measures the particle and the moment when the observer learns the result. This takes only a fraction of second, but it still isn't instantaneous. If the wave function collapses only when it comes into contact with a consciousness, then the idealists will have to postulate that the observer's consciousness emits a signal that travels back through time and tells the apparatus what it should indicate at the precise moment when the particle interacts with the machine. This is a rather odd scenario, to say the least. And it becomes utterly absurd when the observer is replaced by an automatic recording mechanism in which the magnetic tapes are analyzed months after the experiment was over.

M: But if we take the EPR experiment literally, the very coexistence of the measuring apparatus and the mind that conceived it are enough for them to participate in what you call the global wave function. There could be an instantaneous correlation without any information being exchanged.

T: Consciousness only intervenes in the making of the machine and in the interpretation of the results.

M: And that's enough for it to have taken part in the global phenomenon, in such a way that it can't be disassociated from it.

T: The primary role played by the apparatus in fact appears in the interpretation offered by the second group, the materialists. They claim that consciousness plays no part, that the world doesn't depend at all on an observer's presence, and that it would be just the same even if it wasn't observed. As an explanation for the collapse of the wave function, materialists argue that after measurement, the wave function of the "particle plus apparatus" changes so rapidly that only one possibility can materialize. This rapid evolution is supposed to result from the macroscopic nature of the measuring apparatus. But this explanation isn't very convincing and, for the moment, has not been demonstrated rigorously.

Mention should also be made of the theory of "decoherence" developed by the American physicist Wojciech Zurek and others. According to this theory, it is tiny interactions of a physical system with the surrounding environment (for example, a photon or a gas molecule bouncing off that system) that provoke the collapse of its wave function. It is almost as if the environment itself acts as an observer.[4]

M: Nature doesn't measure itself. If a ruler falls beside a plank, this doesn't constitute a measurement. As soon as there's a notion of measurement, we must directly or indirectly introduce the consciousness that planned the measurement, whether the result is immediately perceived or not. Thus consciousness is irreversibly part of the interdependent, global phenomena that we are studying.

Buddhism also says that the observer and the observed can't be separated. They interact and shape each other in a global universe, like two knives sharpening each other. We're structured by our environment, just as we affect our world through our projections, concepts, and habits. Any attempt to pull them apart or to conceive and describe a world totally independent of us is doomed to failure. In the *Avatamsaka Sutra* there are these words, attributed to the Buddha:

There is neither a painting in the mind
Nor a mind in the painting;

And yet, where else can one find a painting
Than in the mind?

Something struck me while you were explaining how Einstein thought that the past, present, and future were all present. This seems to lead to a totally deterministic vision. All we'd need to do would be to read the book of time in order to know the past and the future. It would be pointless to try to change anything, oneself included, since the die would already have been cast, and neither God nor quantum uncertainty play dice.

T: You're absolutely right about that. I don't agree with this deterministic vision, either. In this case, Einstein was the intellectual heir of Newton and Laplace. Newton thought that the universe was a huge machine made up of inert material particles submitted to blind forces. Using a small number of physical laws, the history of an entire universe could be explained and predicted if we managed to characterize it perfectly at a given moment. Laplace summed up this triumphant determinism in his famous declaration:

Consider an intelligence that, at any instant, could have a knowledge of all forces controlling nature, together with the momentary conditions of all the entities of which nature consists. If this intelligence were powerful enough to submit all these data to analysis, it would be able to embrace in a single formula the movements of the largest bodies in the universe and those of the lightest atoms; for it, nothing would be uncertain, the future and the past would be equally present to its eyes.[5]

Time is, in a way, abolished. This inspired Friedrich Hegel's famous remark, "There is never anything new in nature." This sterile, rigid, and dehumanizing determinism dominated until the end of the nineteenth century. In the twentieth century it was swept away by the liberating vision of quantum physics. The role of chance, or what we would call contingency, was recognized in such varied fields as cosmology, astrophysics, geology, biology, and the cognitive sciences. Our world has also been molded by a succession of historical events, such as the asteroid that hit the Earth, causing the disappearance of the dinosaurs and thus giving our mammal ancestors the chance to proliferate.

M: Laplace accounted for such contingent effects in his argument, didn't he? He believed that even they could eventually be explained within his deterministic view.

T: Henri Poincaré, the French mathematician and one of the pioneers of chaos theory, replied as follows to Laplace's deterministic credo:

> *A cause so small as to escape our attention, determines a considerable effect that we cannot help but see. We then say that this was the result of chance. If we knew the laws of Nature exactly and the precise situation of the universe at the initial moment, we could then accurately predict the situation of this same universe at some future moment. But even if the laws of Nature held no more secrets for us, we could have only an approximate knowledge of the initial situation. If this allows us to predict a future situation with the same approximation, then this is all we need. We then say that the event has been predicted and that it is governed by laws. But this is not always the case. It can happen that small differences in the initial conditions create very large ones in the resulting phenomena. A tiny error in the initial state then leads to an enormous error in the final state. Prediction becomes impossible.*[6]

In this way, Poincaré refuted the postulate at the heart of Laplace's argument: that it's possible to know the precise initial conditions of any phenomenon in the universe. From the inevitable large or small inaccuracies of the initial conditions, and the extreme sensitivity of certain systems to their initial conditions, any attempt to predict the future evolution of these systems is doomed to failure.

This is a central tenet of chaos theory, which has become an important complement to physics in the attempt to understand our world. I've often wondered why Newton failed to discover chaos. Uncertainty was lurking at the heart of his equations. Potential unpredictability lay in his theory of gravitation, because it was by using that very theory to study lunar motion that Henri Poincaré discovered chaos. Recently, it has been shown that a slight change in the position or initial speed of a planet such as Pluto could make it slip out of its regular orbit into a chaotic one. The solar system, which was considered to be a well-oiled cosmic machine running on rigid, deterministic laws, is also chaotic. Chaos lurks in the regular, and the unpredictable is never far from the predictable. This chance and indetermination affect not only the planets, stars, and galaxies, but also our everyday life. A simplistic conception of the laws of cause and effect is no longer defendable.

That said, the fact that Newton failed to notice chaos takes nothing away from his genius. On the contrary, his intellectual brilliance in singling out and studying the systems in nature that do evolve in a linear and nonchaotic way allowed him to formulate his monumental theory of gravitation.

In scientific terms, chaos isn't a lack of order, as in the general use of the word. It has more to do with long-term unpredictability. For example, it's impossible to forecast next week's weather, because weather events are extremely sensitive to initial environmental conditions. In order to predict long-term weather, we would need to know those initial conditions with an infinite precision, which is impossible. Even if we were to acquire that perfect knowledge, it would not be possible to communicate it to our computers because of their finite memory. Chaos presents an ineluctable limit on our knowledge. The seeds of ignorance have been planted in the very workings of nature. It would be vain, in an attempt to understand the weather's moods, to set up meteorological stations everywhere. There would still be undetectably tiny atmospheric variations. As they become amplified, these fluctuations can lead to either a storm or a beautiful blue sky. That's why chaos is often explained by what physicists call the "butterfly effect": the flapping of a butterfly's wings in Guyana can trigger a rainstorm in New York. Newton and Laplace's deterministic dream has faded away.[7]

M: The butterfly effect is even clearer when it comes to mental events. A simple thought can lead to planetary convulsions. A feeling of hatred or ambition can set off a world war. Tiny differences in the motivations behind our actions create radically different courses of events, which lead to a vast range of misunderstandings and conflicts.

T: Chaos is at work all the time in our daily lives. You must have experienced occasions when apparently innocent events led to dramatic consequences. An alarm clock fails to go off, so a man misses his interview and the job he wanted. A speck of dust in the gas tank makes a car break down, so a woman misses her plane and escapes death when it crashes into the ocean a few hours later. Insignificant events and imperceptible differences in circumstances can thus radically alter someone's life.

M: A determinist might reply that if it were possible to know the initial conditions perfectly, no matter how subtle they were, and if we had the necessary computing power, we could predict how a series of events would develop.

T: But it's our very inability to know *perfectly* the initial conditions that makes it impossible to predict the future.

M: Yes, absolute accuracy seems impossible because of the subtle impermanence of phenomena, which means that no measurement can be truly instantaneous. A measurement occurs in time, and can't be absolutely accurate because conditions are constantly changing.

T: The uncertainty principle does in fact state that, given that any measurement implies an exchange of energy, it cannot be made in zero time. The shorter the time for the measurement, the more energy is needed. An instantaneous measurement would therefore require infinite energy, which is impossible. So the dream of knowing all the initial conditions with perfect precision is mere delusion. As we've seen, quantum uncertainty means that, in the atomic world, we can't accurately pin down a particle's position and speed at the same time, and so can't trace its trajectory. In the macroscopic world, chaotic phenomena are so dependent on their initial conditions that prediction is ruled out. Liberated from its deterministic straitjacket, nature can give free rein to its creativity. The laws of physics provide the universe with themes for variation and improvisation. By playing with these laws, nature can spontaneously create novelty.

M: Quantum theory has certainly destroyed absolute determinism by bringing probability into the laws of cause and effect. However, when we can't find an event's immediate cause, such as the disintegration of unstable particles or radioactive elements, the idea that the event has therefore happened "by chance" is just one of the possible interpretations of quantum mechanics. By abandoning necessity and adopting chance, physics thinks that creativity has returned to the world. But all we've done is to go from one extreme to the other. Can an event really have no cause? Isn't uncertainty so called because the life of particles doesn't follow a linear form of causality, which is worrying at first sight? Yet an understanding of "pure" chance and probability has been introduced where all we need is interdependence, with its infinite potential for manifestation, in order to explain events. What's more, these notions of creativity and spontaneity are rather reminiscent of the concept of an organizing principle that, whether it be playful or serious, doesn't seem to me to stand up to scrutiny.

T: When I spoke of a playful, creative nature, I was indeed implying the existence of an organizing principle, of the sort envisaged by Spinoza and Einstein. That said, you mustn't think that quantum theory overemphasizes pure chance. There's still a lingering determinism within the theory. Individual quantum events can't be determined, but probabilities for sets of events can be accurately forecast using the laws of statistics. For example, while we can't calculate an electron's precise trajectory, we can calculate the probability of its being at any given point. It's this vestigial determinism that allows our computers and stereos to work. If everything in their electrical circuits was random, then they wouldn't function. As for chaotic phenomena in the solar system, we can't predict the motion of the planets for periods of over several tens of millions of years (i.e., less than 1 percent of the age of the universe). All the same, they have quietly continued orbiting around the Sun for the last four and a half billion years because, although the probability that their orbits will become chaotic isn't zero, it is extremely low. Does Buddhism agree with this notion of bounded unpredictability?

M: In Buddhism, neither pure chance nor necessity can be accepted; they are two extremes, and neither of them stands up to analysis. No effect can be causeless. On the contrary, there are so many causes that it's impossible to come up with a linear, deterministic analysis of causality. Strict determinism holds only where there is a finite number of factors in the cause-and-effect relationship. But, in the global system, there is an undetermined number of elements involved, including consciousness.[8] A system like that necessarily escapes absolute determinism and transcends the powers of discursive thought. Novelty can thus emerge from synergy without having to be explained by a limited number of causes or pure chance, that is to say the absence of any cause.

T: That view is strikingly similar to the chaos theory view.[9] We resort to the notion of chance in physics because we cannot fathom the notion of an infinity of causes. Chaos in physics, as we've seen, isn't the absence of order but the impossibility of making long-term predictions. Referring to this unpredictability as the role of chance is just a shorthand expression.

In fact, what you said about novelty arising from the synergy among certain phenomena is remarkably like the concept of "emergence," which comes out of the new science known as complexity theory, and has become central to physics and biology. For example, one of the great mysteries in modern science is how life arose from matter. How did the inanimate produce the animate? One current idea is that the elementary particles in the primordial terrestrial soup managed to organize themselves into ever more complex states, in a process called "emergence," until they produced the building blocks of life that were capable of reproducing themselves and finally life itself. This form of organization doesn't require outside intervention or mysterious forces. Rather, order "emerges" as soon as the complexity reaches a critical threshold. Biological systems have a hierarchy of organizational levels, and on each new level, new behavior emerges. Thanks to organizing principles emerging on the upper levels, new qualities appear that can't be predicted from the conditions on the lower levels. The behavior of a complex, organized ensemble such as a human being can't be explained in a reductionist way by the behavior of the particles that constitute it. The concept of emergence can be summed up simply as "the whole is more than the sum of its parts."

M: This apparent gulf of existence between a more complex entity and the simpler components from which it is made—this leap from one level of complexity to the next—is a by-product of our inability to appreciate interdependence. Causality is never one-way. If we call "upward causality" the fact that elements on lower organizational levels combine to produce something on a higher level, then we can say that "downward causality" implies that an element on a higher level can influence elements on lower levels. In this way, life influences the planet, social phenomena influence individuals, and consciousness influences our bodies and "our" world. Thus causality isn't simply upward, it's also downward. And it's always mutual. So Buddhism prefers to talk about co-emergence, dependent origination, or reciprocal causality. Consciousness fashions reality and reality fashions consciousness, again like the blades of two knives sharpening each other. A proper understanding of interdependence thus implies transcending the conventional notions of levels of existence or of dualism between "self" and "the world," or between "conscious" and "inanimate."

T: But, by so strongly emphasizing interdependence, don't you open the gates to a new form of determinism? Couldn't someone who knew how a particle was interacting with the entire universe, someone with the supreme knowledge that Buddhism talks about, predict the moment when the particle would disintegrate? In that case we could no longer talk of probabilities.

M: We could suppose that someone who was omniscient—as people are supposed to be who have reached the state of the Buddha—can clearly see the reasons and the implications of each situation. The texts say that only an omniscient mind—which can be compared to a perfect grasp of interdependence—can understand all of the causes and conditions that have brought about the colors of a peacock's tail or the roundness of a pea. But this omniscience in no way implies belief in the sort of determinism involved with such ideas as an organizing principle or a prime cause for everything. As regards matter, interdependence is indeterminate because there is an unlimited number of causes and conditions. As regards consciousness, the concept of freedom of choice is essential, because at each instant we stand at a crossroads.

T: When I mentioned an underlying creative nature in the world that chaos theory and quantum uncertainty have freed from the straitjacket of determinism, I wasn't necessarily implying that there is a consciousness of some kind governing this creativity. I believe that nature evolves and acquires emergent properties according to the laws of organization and the principles of complexity. Can you clarify Buddhism's views on determinism and cause and effect?

M: First, Buddhism rejects the idea that anything can be causeless. If a result could happen without any cause, absolutely anything could lead to absolutely anything else, since what is causeless depends on nothing. So an effect must depend on its causes and conditions. This seems simple, but things become more complicated when we remember that Buddhism also rejects the notion of "objective" reality. The reductionist way of looking at causality supposes that an inherently existing entity with intrinsic properties acts on other entities by altering their properties. But Buddhist logic points out the insurmountable problems that arise when we consider phenomena as concrete, independent entities. So our view of causality is more complicated. In order to truly understand the Buddhist view, we should go through the traditional Buddhist analysis of this problem of causality.

We start with the realization that there can be only four sorts of causality, or means of production, in the world. A thing can be born (1) from itself; (2) from something else; (3) from itself and something else; or (4) neither from itself nor from something else. Then we work our way through the possibilities.

The first step is to acknowledge that a thing can't be born of itself. If it contained all of its own causes, it would then multiply indefinitely without anything being able to stop. When all of the necessary causes are present, the event in question must occur. What's more, if a thing was born of itself, this would mean that it already existed. Production would then be unnecessary. If what had already been born was born again, then the process would never stop.

T: And what about the second possibility, which is more similar to our usual ideas of causality and those of science? Can a thing be produced by "something else"?

M: Buddhism accepts this sort of causality in terms of relative truth. In absolute terms, however, it affirms that if the cause and the effect were totally distinct, then causality couldn't operate. The reasoning goes like this: at the moment when the cause is about to vanish and the effect is about to appear, do the cause and the effect, considered as real, separate entities, have a "point of contact," even for just a fleeting instant?

If yes, the cause and effect exist simultaneously when they are in contact. The effect thus doesn't need to be produced, given that it already exists and the cause is unnecessary. What's more, two simultaneous entities can't work on each other in causal terms, because they can't act on each other in the present instant. (This goes back to what Heisenberg said: "Two simultaneous phenomena cannot be connected by any direct causal action.") On the other hand, if the cause and effect have no point of contact and are totally unconnected, causality breaks down. The two entities have nothing to do with each other and so can't be in a cause-and-effect relationship. What's more, if the cause has nothing to do with its product, anything could be born from anything else. In the words of Chandrakirti:

> *If something could be produced by something intrinsically "other,"*
> *Then darkness could be born of fire*
> *And anything could be born from anything.*

Anything could be born from anything, because if the "cause" entity is "other" in terms of the "effect" entity, all phenomena are equivalent in the sense that they are all "other" in terms of the effect. In that case, any phenomenon could have been the cause. If the cause has already disappeared when the effect appears, then this comes down to saying that the effect happened with no cause and is an *ex nihilo* creation. In other words, if the cause vanishes before the result arrives, then the result never will arrive. A seed can't vanish before giving birth to a shoot. Nor can the cause remain unchanged when the result arrives, just as a seed can't give birth to a shoot without vanishing.[10] To sum up, a concrete, autonomous entity can't produce another one. If the "result" entity already exists at the same time as the "cause" entity, either it doesn't need to be produced, or it takes part in its own production, which is meaningless. If it doesn't exist, its production is impossible, given that a billion causes can never produce something from nothing. Nagarjuna summed up this argument in the following quatrain:

> *If the entity of the effect already exists*
> *What does a cause have to produce?*
> *If the entity of the effect does not exist*
> *How could a cause produce it?*[11]

And Atisha added, in the Torch of the Path to Enlightenment:

> *Something that exists already cannot, logically, be born.*
> *Just like nonexistent things—which are like flowers in the sky.*

So this is Buddhism's conclusion: What seems to us to be a cause-and-effect relationship can only be possible if *neither the cause nor the effect* exists independently and permanently. We thus come back to the phrase "because everything is emptiness, everything can exist." The nonreality of phenomena is the precondition for their appearance. These "simple appearances" then evolve according to a law of causality based on interdependent phenomena with no inherent existence. To quote Nagarjuna once more:

> *There is not the slightest thing*
> *That does not come from a dependent origin.*
> *And therefore there is not the slightest thing*
> *That is not emptiness.*

Modifications of these interactions bring about the chain of cause and effect, without its being necessary to postulate the existence of separate entities, each containing all of its own properties—what physicists call "local" properties.

T: That's right. Einstein, who rejected any idea of a global, interdependent reality, called them "hidden local variables." But experiments on the EPR phenomenon have shown that these local variables don't exist.

M: Let's now turn to the last two possibilities. Something can't be born both of itself and of something else for the same reasons as in the preceding arguments. So, can something be born neither of itself nor of anything else? No, it can't. For if it could be born with no cause, anything could be born anytime, anywhere, and anyhow.

T: So that leaves us with what?

M: The only solution is interdependence, a co-production in which phenomena condition one another mutually within an infinite network of dynamic, impermanent causality, which is incomprehensible to a linear way of thinking, which is innovative without being arbitrary, and which eludes the two extremes of chance and determinism. To sum up, an inherently existing object can't have a cause and can't depend on anything else. If everything existed in this fashion, nothing would come about, causality wouldn't operate, and the world of phenomena would be permanently frozen. The fact that things seem to happen in the world of appearances, or relative truth, is possible only because cause and effect have no intrinsic existence. It is said in Prajnaparamita's Transcendent Knowledge:

> *They have no ending and they have no origin;*
> *They are not nothing, nor are everlasting;*
> *They do not come, they do not go;*
> *They are not one, they are not more than one.*[12]

A correct understanding of emptiness thus stops us from falling into the traps of realism or nihilism. Meditation on emptiness attenuates the belief in the real existence of things. But you mustn't become attached to emptiness as a belief. If you do, you will relapse into nothingness. In the Garland of Jewels, Nagarjuna writes, "Since we find nothing real, how can we find something unreal? Indeed, the 'nonexistent' can only be conceived of in relation to what is existent." In his Fundamental Treatise on Wisdom, he concludes:

When emptiness is wrongly understood,
It leads the ignorant to their perdition.
And thus in neither "is" nor "is not"
Does the sage abide.

According to Buddhism, the secret of understanding reality lies in the union of emptiness and appearances. When things are empty, they appear; when they appear, they are empty. Over and above the limitations of simple theoretical rationality, a true understanding of this statement can only be reached by means of direct contemplation. As it is said in Transcendent Knowledge:

People say, "I see a space"—
But how can space be seen? Examine what this means.
In such a way the Buddha spoke of "seeing" the ultimate nature of things;
He found no other word than "seeing" to express himself.

THE VIRTUAL FRONTIER

10

A MIND-BODY DUALITY?

What exactly is consciousness, and where does it come from? Did this ineffable phenomenon arise naturally through the processes of evolution, once brains became sufficiently complex? Or has conscioucness coexisted with the material universe through all time? Does conciousness depend upon a brain to generate it? Or does it exist apart from any material embodiment? Does the idea of a division between the mind and the body make any sense? If consciousness can exist apart from a body, does this perhaps explain accounts of reincarnation?

THUAN: Most biologists believe that in the process of evolution, consciousness arose when the networks of brain cells in living beings reached a threshold of complexity. This theory implies that consciousness emerged, just as life itself, from inanimate matter.[1] Does Buddhism agree with this view?

MATTHIEU: Buddhism agrees with most of the opinions of science concerning the universe's evolution—except, of course, for the notion of a "beginning"—but it disagrees about the origin of consciousness.

According to Buddhism, consciousness, just like all phenomena we perceive, has no intrinsic reality. Consciousness as we experience it in everyday life belongs to the realm of relative truth. It is just a helpful concept. That said, Buddhism believes in a difference between a "conscious unreality" (the mind defined as a stream of conscious instants) and "unconscious unreality" (the material world that it perceives). We do not believe, as so many biologists argue, that consciousness, or the mind, arises out of the matter of the brain.

Buddhism distinguishes three levels of consciousness: gross, subtle, and extremely subtle. The first of these is the level of the biochemical workings of the brain. The second is the subjective experience that we customarily call consciousness, that is to say the mind's faculty to, among other things, examine itself, to ponder its own nature and exercise freedom of choice. The third level, that of the extremely subtle, which is the most important, is also called "the fundamental luminosity of the mind." This is a state of pure awareness that transcends the perception of a subject/object duality in the world and breaks free from the constraints and traps of discursive thought.

These three types are not separate streams of consciousness, but lie at different, increasingly deep levels. The gross and subtle levels both arise from the fundamental level, as opposed to the other way around, as might be expected. The brain, and in fact the whole body, even extending outside the body to the environment, provide gross and subtle consciousness with the conditions that allow them to manifest themselves. These levels of consciousness are both shaped and modified by the brain and environment, and can in turn modify the brain and body. The activities of these levels of consciousness are correlated to the brain, and they can't manifest themselves without a body.

Fundamental consciousness is quite different. In the tantras—Buddhism's profoundest vision—fundamental consciousness is called "pure awareness" *(rigpa)*. This type of consciousness is not dependent upon the workings of the brain. It is free of confusion and transcends discursive thought, both positive and negative, as well as the error of mind that is called samsara and the elimination of error that is called nirvana. It's also called "the primordial continuity of the mind," "natural luminosity," "the ultimate nature of the mind," "essence of Buddhahood," "the natural state of consciousness," "unique essential simplicity," "primordial purity," "spontaneous presence," and "absolute space."

This pure awareness can make itself manifest without the need of the trappings of the brain. We say that this primordial consciousness has a natural "creativity" *(tsel)* in the form of various thoughts that constitute the "play *(rolpa)* of pure awareness." If one recognizes that thoughts thus arise out of pure awareness, one's understanding of awareness is enhanced rather than obscured by these thoughts. In that case, we say that thoughts become "ornaments" *(guien)* to pure awareness. In normal life, only the gross and subtle levels of consciousness can be discerned, because the fundamental level has been cloaked by the veil of ignorance, just as the sun can be momentarily covered by clouds. However, ignorance can no more affect the primordial nature of consciousness than clouds can affect the sun.

T: But where do conscious phenomena come from? Do we need a "spark" to set alight life and consciousness in inanimate atoms?

M: This "spark" idea creates a major problem. According to this conception, consciousness had a beginning. If so, either it was created *ex nihilo* (without a cause, or by a Creator—and we've already seen how Buddhism refutes these two ideas), or it gradually came to life in inanimate matter, as most biologists and physicists think. One of them, the physicist Brian Greene, wrote the following letter to me: "I think that consciousness is a reflection of microphysical processes (of great and stupendous speed and complexity). Although the qualitative features of consciousness differ dramatically from the properties of the physical constituents in which it is based, I do not think this points toward there being something 'else' than the physical structure."

Buddhism would answer that cause and effect must have a common nature, when the cause is substantially responsible for the effect. (When the cause is simply a cooperative condition of the effect, as described below, the two may be quite dissimilar.) A moment of consciousness can only be caused by a preceding moment of consciousness. If something could be born from something utterly different, then anything could be born from anything else. Thus the fundamental level of consciousness cannot have arisen from inanimate matter, and it doesn't necessarily, always and in all contexts, depend on being embodied in a physical form.

The Dalai Lama explained this idea as follows:

> *It's clear that consciousness depends on the functioning of the brain, so there is a causal relationship between brain function and the arising of gross consciousness. But here is a question I continue to consider: What* type *of causal connection is it? In Buddhism we speak of two types of causes. The first is a* substantial cause, *in which the stuff of the cause actually transforms into the stuff of the effect.*[2] *The second is a* cooperative condition, *in which one event takes place as a result of a preceding event, but there's no transformation of the former into the latter. . . . Let's apply this to the causal origination of consciousness and its relationship to brain function. What type of causality exists there? We have, experientially, two types of phenomena that seem to be qualitatively distinct: physical and mental phenomena. Physical phenomena seem to have a location in space, and they lend themselves to quantitative measurement, as well as having other qualities. Mental phenomena, in contrast, do not evidently have a location in space, nor do they lend themselves to quantitative measurement, for they are of the nature of simple experience. It seems that we're dealing with two very different types of phenomena. In this case, if a physical phenomenon were to act as a substantial cause for a mental phenomenon, there would seem to be a certain lack of accord between the two.*[3]

Here is a simple image that illustrates the Dalai Lama's point: the seed is the substantial cause of the flower, while the sun and water are cooperative conditions.

T: The materialistic, or "monist," position is that the brain is comprised of a mass of neurons, and that consciousness is simply the result of the electric currents that run through the neuronal circuits. As the eighteenth-century doctor Pierre Cabanis expressed this idea, "the brain secretes thoughts as the liver secretes bile."

Neurobiologists argue that our brains, and in turn our minds, are constantly being shaped by our interactions with the world around us. Consciousness is born from this constant interaction.[4] The meaning of the world emerges from the permanent activity of our bodies in a particular environment. Neurobiologists also argue that consciousness emerges from the inanimate matter that constitutes our brains. They see no need for an additional ingredient that transcends the physical.

When Buddhism conjectures that there is a level of consciousness that transcends the physical, isn't it falling back into Descartes's mind-body dualism, in which there are two distinct types of reality, that of the mind (or thought) and that of the material world? According to Descartes, the mind is pure consciousness, doesn't occupy space, and can't be subdivided. On the other hand, matter is unconscious, occupies space, and can be divided. Man has a double nature: we think, but we also have a material body.

M: Buddhism's conception is radically different from Cartesian dualism. We believe that there's merely a conventional difference between matter and consciousness because, in the end, neither of them has an inherent existence. Because Buddhism refutes the ultimate reality of phenomena, it also refutes the idea that consciousness is independent and exists inherently.

One of the arguments we use to convey this point depends on the ability of subtle consciousness to be aware of itself. We would argue that a truly existing consciousness could no more reflect on itself as an object than a sword can cut itself. One might refute this point by saying that consciousness can shed light on itself like a flame, but Buddhism would reply that it's the nature of a flame to burn, and it doesn't need to shed light on itself. If a flame could light itself, then shadows could also create their own darkness. This reasoning applies, however, only to the concept of consciousness as an intrinsically existing function. Considered as the continuum of a "cognitive function," consciousness can indeed know itself.

When we realize that we are thinking about something, a flower for instance, our attention is consumed by the mental picture formed by our perception of the flower. Yet in order to know that it is experiencing the thought of a flower, consciousness must be able to know itself. This is possible because the primordial aspect of consciousness, what we've called "fundamental luminosity," has a natural self-conscious quality that lies beyond the subject-object duality. We thus escape from the infinite regress of another observer observing the observed. One of the qualities of enlightenment is the ability to stay in this state of nondual "enlightened presence," which is characterized by a direct knowledge of consciousness's "luminous" nature and doesn't involve mental pictures. This fundamental level of consciousness and the world of apparent phenomena are linked by interdependence, and together they form our perceived world, the one we experience in our lives.

Descartes's dualism, lacking as it does the concept of interdependence, is limited by an absurd notion of the strict wall between mind and matter. How can consciousness interface with the material world if both exist as such independent, inherently existent entities that have nothing in common?

T: According to Descartes, this interface takes place in the tiny structure in the brain known as the pineal gland. He argued that through this gland, the mind reacts to the body's passions and moods. But the mind retains the ability to separate itself from "low" compulsions such as desire or hatred, and to work independently from the body. As for the body, Descartes saw it as a perfect machine, governed by the laws of physics.

I don't need to tell you that the idea that the pineal gland is the seat of consciousness was refuted by science long ago. But the question that vexed Descartes—how to identify the mind within the body—has still not been definitively answered. Critics of Cartesian dualism referred to his notion of the mind derisively as "the ghost in the machine."

In trying to solve the question of how the mind relates to the physical world, some people have suggested that quantum uncertainty provides an explanation for how consciousness can interact with matter. But others argue that the notion that an immaterial phenomenon that cannot be quantified could interact with a material system is incompatible with the law of the conservation of energy (nothing is created and nothing lost), which is one of physics' sacred principles, and which demands that all components of a system be rigorously accounted for.

M: But this seems incompatible with the realization that matter is not "solid" in the way that had been thought, and that in fact matter and energy are interchangeable. Buddhism rejects the distinction between the material and immaterial. We say that "emptiness is form and form emptiness." The dichotomy of material and immaterial makes no sense. Even if there are qualitative differences between animate consciousness and inanimate matter, there is no basic incompatibility that outlaws interaction between the material world and consciousness. Interdependence provides an interface.[5]

T: In other words, Buddhism says that the distinction between the interior world of thought and exterior physical reality is artificial. The antithesis between internal and external realities is a mere illusion. There's only one reality.

M: Or rather only one unreality! In this sense, the mind/matter dichotomy turns out to be another example of our attachment to solid reality and is just a concept.[6]

T: Cartesian duality is also based on the idea that the physical world lacks any subjective qualities. But, as we've discussed before, quantum mechanics shows that the role of an observer is part of an interdependent process that produces an observable phenomenon. When you mentioned the complementarity between the brain and consciousness, I couldn't help thinking of Niels Bohr's famous principle of complementarity. I think that the mind complements matter, just as the "particle" aspect of matter complements its "wave" aspect.

M: What's more, the concept of the dichotomy between our inner self and the outside world leads to our inability to perceive the true nature of reality. This imaginary line in the sand that we draw between interior and the exterior, consciousness and matter, self and other, gives rise to the ego. Given that this distinction is fictional, we clearly can't satisfy all of its whims. By trying to do so, we constantly find ourselves at loggerheads with reality. Chandrakirti summed up the resulting frustration like this:

First conceiving an "I," we cling to an ego.
Then conceiving a "mine," we cling to a material world.
Like water in a water-wheel, helplessly we circle;
I bow down to the compassion that arises for all beings.[7]

T: The artificial division between the self and the exterior world was also pointed out by the founders of quantum physics, such as Schrödinger, who wrote, "Subject and object are only one. The barrier between them cannot be said to have been broken down as a result of recent experience in the physical sciences, for this barrier does not exist."[8]

A much more promising approach to understanding how consciousness arises out of the inanimate world comes from the new science of complexity. According to this view, consciousness "emerges" once the networks of neurons in the brain reach a critical threshold of complexity. This view is based on the observation of certain so-called "open" physical and chemical systems, which are ones that interact with their surroundings. This interaction makes them reach "bifurcation points" where they suddenly become far more complex, or, in other words, far more organized. This process can be seen in the case of boiling water. Below a certain temperature, water remains in a homogeneous condition; there is no shape to the liquid, or, as physicists would say, no structure. As the temperature increases, no change is apparent until the water abruptly displays a highly organized bubbling of convection cells, when it has reached the critical temperature. The water has bifurcated from an unorganized state to an organized one. Many biologists now believe that evolution happened in a similar way, progressing from one bifurcation to the next, becoming increasingly organized in sudden jumps, thereby ascending the scale of complexity from the inanimate to the animate.

The driving forces behind the bifurcations were environmental events that knocked the biosphere out of equilibrium, such as changes in temperature or levels of oxygen in the atmosphere. No extra ingredient is needed, contrary to vitalist theories, such as the one defended by the French philosopher Henri Bergson, who thought that there was an *élan vital* that pushed biological systems into becoming more organized and developing in an effective and creative way. One of the striking discoveries of complexity is that conditions of nonequilibrium can produce organization.

There are several reasons why this is a plausible scenario for the evolution of consciousness. First, living organisms are prime examples of open systems. Life can't exist in isolation; it's constantly exchanging energy with its environment, either by feeding or expelling its waste. Also, there are almost constant destabilizing events in the environment, which break the equilibrium of the biosphere and knock it out of balance. These changes can be either gradual or sudden. The progressive oxygen enrichment of the Earth's atmosphere by plant life is an example of gradual change. Examples of sudden change are solar eruptions, which thrust streams of energy-filled particles toward Earth. Another example is the huge asteroid that hit our planet 65 million years ago and which, as we've already said, caused the demise of the dinosaurs, as well as of three-quarters of the other animal and vegetable species of the time.

According to this theory, evolution should happen not gradually but in sudden jumps. Paleontology seems to lend support to this theory, rather than to Darwin's cherished idea of a gradual evolution. If evolution had been continuous, then we should be able to find all of the intermediary forms between the main groups of living beings. But this isn't the case. Some biologists, such as Stephen Jay Gould and Niles Eldredge, think that biological evolution happens through a series of jumps, in successive stages of "punctuated equilibrium." Living species remain unchanged for a long time, but then undergo radical changes in a relatively short period. Like the "quantum leaps" of atomic physics, evolution proceeds in "evolutionary leaps." We can imagine that it was during one of these leaps that the sparks of life and then consciousness appeared.[9]

M: If this emergence process allows for interactions between the inanimate and the animate that go both ways—both ascending and descending on the scale of complexity—then it accords well with the Buddhist vision of the interface between the body and consciousness. On the ascendant side, the environment and the body influence mental events, though we would not say that they *produce* consciousness. On the descendant side, consciousness influences the body (it has been shown, for example, that the expression of certain genes is switched off in children who lack affection[10]). Consciousness also shapes what we perceive as "our" world. The world we perceive isn't an illusory projection generated by a truly existing mind (as idealists would have it), but rather has been fashioned by the mind, as a vase is thrown by a potter. This fashioning is achieved by the tendencies that consciousness have accumulated during countless existences. The shared experiences of consciousness of similar kinds are what we call "collective karma," and explain why we all see the world in a similar way, whereas our very different individual experiences are what we call "individual karma."

T: Causality can indeed work in both directions according to the emergence view. Not only do the lower levels create the higher levels, but the upper levels influence the lower ones. In this way, consciousness acts on the body. In fact, individual consciousness isn't at the tip of the pyramid. Farther up, we have collective consciousness, which emerges from the shared experience of living together in society, and results in culture and religion. This is what has been called the "social-cultural-historical" factor that models literary, artistic, and scientific work as well as social and political institutions. As an example of downward causality, a change in government can lead to new economic and social policy that can affect each citizen's mental state.

M: I should reiterate, however, that Buddhism distinguishes between levels of consciousness. We can agree that the emergent properties of matter provide the conditions for the working of gross consciousness. "Emergence does not deal with the substance of components," says Francisco Varela, "but with their pattern of relationship, which is in itself immaterial but not disconnected with their physical basis." [11] But from a Buddhist point of view gross consciousness is not *created* by the matter of the brain, it rather arises from extremely subtle consciousness, which itself is made of a beginning-less succession of moments of consciousness.

T: But then how does Buddhism understand the transition between consciousness and matter?

M: Within the enlightened mind, the state of Buddhahood, one can distinguish five wisdoms, which are purified aspects of five afflicting mental factors (hatred, desire, ignorance, pride, and jealousy). At the level of the extremely subtle consciousness, these five wisdoms are expressed as five luminous aspects, which are the natural radiance of Buddhahood, symbolized by five colors—yellow, white, red, green, and blue. These five luminous aspects are themselves expressed as five energies, which manifest on a grosser level as the five exterior elements—earth, water, fire, wind, and space—and the five corresponding interior elements in the body—bone and flesh, blood and humors, vital heat, breath, and cavities.

These energies are the moving force of consciousness. One sometimes compares them to a blind horse and the consciousness to a crippled rider. Without the rider, the horse does not know where to go, and without the horse, the rider can't move.

T: If I understand you correctly, to avoid a discontinuity between the inanimate and the animate, Buddhism envisages a continuous stream of consciousness that goes from one physical embodiment to the next.

M: Yes, and although we tend to emphasize the continuity, and interactions, between consciousness and the body, and think of consciousness as what the neuroscientist Francisco Varela calls the "embodied mind," Buddhism believes that consciousness doesn't necessarily need a physical embodiment. According to Buddhism, it can, for a time, experience a world that is "formless," in which there are no physical manifestations. In the intermediary state between death and rebirth as well, which we call *bardo,* there is perception of forms and of a mental body, but no physical framework. This possibility, which most biologists find unacceptable, is the main difference between Buddhism and the natural sciences.

T: If there were a consciousness without a physical framework, what sort of relationship could it have with the material world?

M: A relationship that is part of the interdependence of the global world. Even without a physical form, consciousness still isn't "disconnected" from the world of phenomena.

T: How does Buddhism defend this thesis?

M: In two ways. First, we point to the cases of people who have experienced the intermediary stage between death and rebirth—or *bardo.* In the West, near-death experiences (or NDEs) are well documented.[12] There are many statements taken of people who have been clinically dead, even for a very short time, and then have been reanimated. Generally, they speak of intense joy, universal love, of traveling toward a dazzling light at the end of a dark tunnel, and a point of no return where they must choose between going on and coming back to life. Witnesses generally talk of feeling reluctant about returning to their physical bodies. They sometimes mention terrifying experiences that are reminiscent of how we might imagine Hell. People who have gone through such experiences generally come back transformed and decide to live their lives in a different way.

Buddhists argue that those who have been through an NDE haven't yet crossed the threshold of death. Yet others (called *delok* in Tibetan), who are generally advanced in the practice of meditation, or else people who have lived out similar experiences for a long period, can describe in great detail the various stages of death and the intermediate stage, or *bardo,* between death and rebirth.

T: I've read a few books about NDE. What struck me in them was the extraordinary similarity in ways that different people describe that intermediate stage between life and death, as a profound sensation of peace, compassion, a powerful light, and so on. Particularly astonishing to me is the fact that some patients, when they wake up, can describe what's been happening in the room. They claim that their spirit "left" their body and observed what they shouldn't have been able to observe, given that they were clinically dead.

M: The other way we defend the idea that consciousness can exist without a body is by referring to the phenomenon of previous lives. Many people have conjectured that they have had prior lives, but a lack of scientific study means that these claims are generally not accepted. There are a few exceptions, however, where we can reasonably exclude deception and pure coincidence.

One of these is the case of Shanti Devi, was born in Delhi, India, in 1926. When she was about four, she started saying strange things to her parents. She told them that her real home was in the town of Mathura, where her husband lived. At first they were amused, but soon started to worry about her sanity. But Shanti Devi was also intelligent and a nice child. However, she kept saying the same thing for two years, which finally got on their nerves. At the age of six, she ran away from home and vainly tried to walk to Mathura, which is over three hundred miles from Delhi. One day she told one of her school friends that her name wasn't Shanti Devi but Lugdi Devi, and that she was married and had had a child whom she hadn't been able to take care of because she'd died ten days after giving birth. Everybody at school made fun of her. She burst into tears and ran away. She wandered around in desperation for some time before stopping near a temple. She then told the whole story to a woman who tried to comfort her. Back home, everyone was panicking. Her father went out to look for her and finally found her. This time he seemed shaken by his daughter's determination. But, during the next two years, nothing happened. Shanti Devi retreated into herself.

Eventually, however, her teacher and headmaster became so intrigued by the studious, serious girl that they went to see her parents to try to clarify the whole affair. They questioned her for a long time, and she answered them confidently. She described her old life in Mathura with her merchant husband, and claimed that she could easily recognize people and places. During this conversation, she continuously used words from the Mathura dialect, which nobody in her family or school spoke. The teachers pressed her for her husband's name. But, in India, it's considered indecent for a wife to name her husband. Shanti Devi hesitated, hid her face in her hands, then shamefully muttered "Kedar Nath." Going against the objections of her parents, who would have preferred to forget the whole story, the headmaster made inquiries in Mathura. Sure enough, he found a merchant whose name was Kedar Nath. The headmaster wrote to him and, a few weeks later, got a reply.

The astonished merchant confirmed that nine years before, his wife had indeed died ten days after giving birth to their son. He obviously wanted more information, but cautiously started by sending one of his cousins to Delhi. The little girl immediately recognized this man, whom she'd never seen, welcomed him warmly, told him that he'd put on weight, that she was sad to see him still unmarried, and then asked him all sorts of questions. This cousin, who'd come thinking that he was going to unmask an impostor, was flabbergasted. He also started asking her questions, but soon told her to shut up when she told him how he had courted her when her husband had been away. The cousin then exclaimed, "Lugdi Devi was the most wonderful woman in the world, she was a saint!" She then asked for him news of her son.

When he heard all this, Kedar Nath nearly fainted. He decided to go to Delhi with his son, with the intention of passing himself off as his brother. But no sooner had he introduced himself under his false name than Shanti Devi exclaimed, "You're not my *jeth* ["brother-in-law" in the dialect of Mathura], you're my husband, Kedar Nath." Then she rushed into his arms in tears. When the son, who was just slightly older than the little girl, came into the room, she kissed him as a mother does. All of the witnesses of this scene were astonished. The discussion grew more and more detailed. Shanti Devi asked Kedar Nath if he'd kept the promise that he'd made on her deathbed not to remarry. She then forgave him when he admitted having taken a second wife. Kedar Nath stayed several days in Delhi and asked Shanti Devi a thousand questions that she answered with quite disconcerting accuracy. He left convinced that she was definitely the reincarnation of his wife.

But things didn't stop there. Word got around, and to everybody's surprise, Mahatma Gandhi himself came to see the little girl. He was fascinated by her. Shanti Devi told him, among other things, that Lugdi Devi had been very religious. Gandhi stroked the little girl's hair and said, "I hope to hear more from you when you're in Mathura. My thoughts will be there with you. You need the truth. Never stray from the path of truth, whatever the cost." He then sent her to Mathura, with her parents, three respectable townsmen, and some lawyers, journalists, and businessmen, all of high intellectual repute. On November 15, 1935, this party arrived at Mathura station. A crowd was waiting for them on the platform. At once, the child astonished everyone by recognizing the members of her "former family." She ran over toward an old man and cried out "Grandfather!" and asked him for news of her sacred basil. The old man was astonished. Just before dying, Lugdi Devi had given him her sacred basil, a plant revered in India for its spiritual and medicinal value.

Then she led the procession straight to her house. In the next few days she recognized dozens of people and places. She met her former parents, who were overwhelmed. Her current parents were extremely worried that she wouldn't stay with them. Despite being torn, she decided to go back to Delhi with them. Thanks to her questions, she'd found out that her husband had kept none of the promises that he'd made to her on her deathbed. He hadn't even offered to Krishna her savings of 150 rupees that she'd hidden under the floorboards for the salvation of her soul. Only Lugdi Devi and her husband knew of this hiding place. Shanti Devi forgave her husband for all his failings, while everyone who heard her admired her more and more. The commission of local worthies carried out its investigations scrupulously, cross-checking information and accumulating details. It concluded that Shanti Devi was indeed the reincarnation of Lugdi Devi.

Shanti Devi then lived a modest life and remained single, because she'd promised her husband not to marry in a future life. She never tried to benefit from her fame, and after studying literature and philosophy, she gave herself over to prayer and meditation. At the end of the 1950s, she agreed to tell her story over again.[13]

T: That really is an astonishing and striking story.

M: It's an exceptional case, but by no means an isolated one. Ian Stevenson, a professor at the University of Virginia, has studied statements made by a few hundred people who claim to have similar recollections. He has picked out twenty cases in which the detail of recollections is hard to explain as anything other than the product of actual memory from previous lives.[14] They are always ordinary children.

T: Stevenson teaches at the same university as I do, and I've discussed his work on memories of past lives with him. He says that it's a hard subject to study because most cases turn out to be frauds. If it is possible to remember past existences, why can so few people do so?

M: When you wake up in the middle of the night, after a general anaesthetic or a fainting fit, you feel extremely confused and for a few moments don't know where you are. This transient interruption of our mental faculties caused by minor traumas is similar to what happens after death, but not to the same degree. It's easy to see that death is far more traumatic and so we forget more. If, however, we possess a great clarity of mind when we die, or if we die young, then memories can go over into the next life. This phenomenon occurs during early childhood because, as we get older, our new life imposes itself on our consciousness, and our impressions of our past life disappear. The obscurity caused by death is less marked in people who have reached an advanced stage of contemplative mastery in their previous life and know how to pass lucidly through the intermediate stage between death and rebirth. That's why in Tibet we think these kinds of memories are most commonly found in young children who are reincarnations of dead sages.

T: Such as the case of the Dalai Lama, who is generally chosen on the basis of a young child's memories. But this may not be an adequate answer, because in the cases of the twenty children confirmed by Stevenson, none of them recalled being spiritually advanced in any way. Apparently none of them was a sage with a high level of spirituality in his or her past life. You have some personal experience with at least one person who is considered to be the reincarnation of a sage. In *The Monk and the Philosopher,* the dialogue you published with your father, you mention the reincarnation of your teacher, Khyentse Rinpoche.[15]

M: That's right. I considered that I could describe this case, because it's the only one I've witnessed directly. Here I shall mention one event among others. A Tibetan master, who lived in the mountains in Nepal, had identified the child in dreams and visions, and decided to conduct a longevity ceremony for the young incarnation. And so about a hundred of Khyentse Rinpoche's disciples met at a sacred site in the east of Nepal. On the last day, there was a special ritual, during which the officiant handed out a consecrated substance.

Now, when the child saw that the master was about to start, he decided to distribute it himself, though he was only two and a half at the time. Very calmly he called over his mother and gave her a drop of the substance, then, in the same way, summoned Khyentse Rinpoche's grandson and twenty other people. After having blessed those around him, he was asked by a monk, "Well, have you finished?" The child answered, "No, no." Then he pointed at someone in the crowd. Among the hundred or so people present was a group of Bhutanese that had just arrived from the Nepalese border, which is three days' walk away. One of them was an old servant of Khyentse Rinpoche.

Another monk went over to indicate various people in the crowd in the direction the child was pointing—"This one? Or that one?"—and so on, until he reached Khyentse Rinpoche's old Bhutanese servant. The child then cried out, "Yes, him!" So the old man, who had burst into tears, was brought over to receive the child's blessing. This event is particularly important for me, because I witnessed it. Subsequently the child recognized in a quite astonishing way other people who had been close to Khyentse Rinpoche.

The Dalai Lama, who is well known for his simplicity and modesty, says that he can't remember his past lives. He does say, however, that when he went into the thirteenth Dalai Lama's room for the first time, he pointed at a bedside table and asked for his teeth. And, sure enough, his predecessor's false teeth were in the drawer! There are countless similar examples in Tibet, at different times and places, and it seems hardly likely that they're all either frauds or coincidences.

A theory can be disproved by just one exception. For example, the theory "all swans are white" can be based on thousands of sightings and still not be absolutely certain. It can then be destroyed by the appearance of just one black swan. Denying the truth of past and future existences is thus quite different from refuting, for example, the existence of permanent autonomous entities in the world of phenomena. This is based on solid logic. If you wanted absolutely to disprove the existence of successive lives, then you would not only have to refute all the existing reports, but also demonstrate that reincarnation is impossible.

The idea that we can be reborn many times is totally alien to Western culture, and so the very mention of such eyewitness accounts is seen as a provocation, and they're often thrown out indignantly. Such rejection results from an instinctive repulsion for questioning our deeply ingrained metaphysical opinions. Personally, I have no desire to impose such reports as facts. All I want is for the question to be examined more rigorously and collectedly.

T: The cultural effect explains why most reports of memories of past lives come from countries where people believe in rebirth. If the incidents you mentioned happened in the West, they'd be dismissed as either childish antics or signs of mental disturbance. I agree that this question should be studied with as much scientific rigor as possible. I must admit that memories of past lives would be rather handy in helping us progress in this life! We could take advantage of everything we experienced in previous lives and so develop more harmoniously. Our understanding of good and evil would be greater.

M: That is in fact one of the aims of spiritual transformation. If we acquire a certain maturity in this life, even if we don't remember it during the next life, we aren't starting totally from scratch and we ascend the steps of spirituality more easily thanks to what we achieved before. A musician who stops playing the piano for several years will have clumsy fingers when he starts again, but will soon regain his virtuosity.

T: Past lives might also explain child prodigies, such as Mozart, or gifts in general. Did Einstein's prodigious insights into physics come from a long reflection during his previous existences? More generally, could all intuitions be memories of past lives? This would also explain feelings of déjà vu, which come upon us in certain places where we've never been before, or with certain people we haven't met before. In fact, doesn't Buddhism claim that the important people in our present lives (for either good or bad reasons) interacted with us in past lives? Isn't their karma linked with ours?

M: According to Buddhism, all beings have been linked to us at one moment or another since time without beginning. They've all been our fathers, mothers, friends, and enemies. But, of course, there are beings with whom we've established stronger links that can be carried over from one life to the next.

T: Does Buddhism believe, then, in a collective consciousness that pervades the world?

M: We don't envision a global consciousness that's common to all beings and can be found in all phenomena, but rather individual continua of consciousness that go from one existence to the next. These continua can perhaps best be compared to waves in the ocean. When we watch waves rolling onto a beach, it appears that large amounts of water are moving forward to the beach. But this isn't the case. The particles in the water that are forming the waves are actually moving in circles as the energy that creates the swells goes by. The particles don't travel toward the beach.

T: That's why a bottle in the sea isn't carried away by a passing wave. All it does is bob up from the bottom to the crest of the wave. A wave moves on, but without actually carrying the water that seems to make it.

M: The transmission of consciousness from one state to another can also be compared to the transmission of knowledge. During a lesson, there is certainly a passing of knowledge, but this knowledge cannot be called a "thing"; it doesn't literally move from one mind to the other. We can also think of the type of transfer achieved by the use of a mold—a shape is reproduced, but no substance is transferred.

When a moment of consciousness passes, a new one arises identical to it in nature, which is mere cognition but varying in "color," according to its contents. There is simply a continuum of interlinked moments, but there is no underlying entity that endures as the "experiencer" of a stream of events.

The succession of states into which a consciousness passes—and I should point out at once that the words "reincarnation" and "rebirth" are just approximations of this experience—are comparable, to a certain degree, to a something like a radio wave, which transmits information but without itself being concrete. An individual's future lies in the transformations of this wave. The nature of our actions and thoughts determines the states associated with our consciousness.

A physical wave can be destructive, like a radioactive discharge for example, or the source of well-being, such as sun rays warming up a weary traveler. A radio wave can launch an appeal for war or for peace. In a similar way, the modifications made to our wave of consciousness by our thoughts and by the altruistic or malevolent motives behind our words and actions are expressed as happiness or suffering.

The wave of our consciousness continuum contains all of our experiences in this life and in our past lives in an infinitely complex web of positive and negative elements, and moments of lucidity or confusion. Consciousness can be either purified or darkened.

There are constraints, however, on the degree to which we can change our consciousness, owing to the entrenched habits of thought and emotion we have acquired. Our consciousness may have developed "bad folds," and we need spiritual training to undo this sort of conditioning. The ultimate state of purification is the enlightenment or Buddhahood to which every practicing Buddhist aspires.

T: So you're saying that there's a consciousness wave, so to speak, associated with each person?

M: In the same way that a given wave rolling onto a beach is distinct from all the others before and after, an individual consciousness, or what we would call a person, is distinct. But we do not believe that there must be an "ego" that travels with the wave of consciousness.

T: If there's no "ego" associated with our stream of consciousness, how can memories become attached to it? Isn't the idea of an "ego" essential to the idea of memory? Our conception of ourselves is to a large extent determined by our past experiences. And this is why we have a feeling of personal identity.

M: If the memory was dependent on an "ego," then those who freed themselves from the sense of having an ego would become amnesic! We must avoid confusing the conceptual notion of an "ego" with the stream of individual consciousness. The lack of an "ego" doesn't stop the workings of a memory that is imprinted in the cerebral system and that modifies its own gross consciousness. It does not stop either memories or tendencies being associated with the subtle aspect of the continuum of consciousness. We don't need to imagine an "ego" lording it over these processes. The "ego" is a label attached to what are technically termed our "psychophysical aggregates," meaning the collective of physical and mental phenomena that generate the sense of self in our minds.

T: How can you reconcile this concept of a stream of consciousness with the neurophysiological evidence that memories—which are so essential to generating this sense of self—are created by neural networks in the brain?

M: Certainly there is a close relationship between the neural workings of the brain and the gross aspect of consciousness. That is why the brain's physical health or sickness can so profoundly affect this type of consciousness. But remember that we argue that gross consciousness is just a manifestation of the more fundamental level of consciousness, the extremely subtle level. We believe that the continuum of this subtle consciousness can carry memories, just as a wave can carry information.

Because of entrenched beliefs, this concept of levels of consciousness is extremely difficult for most scientists, and indeed for most people in general, to accept. For an experienced contemplative mind, however, these levels can be experienced. Francisco Varela wrote as follows on this subject:

> *These subtle levels of consciousness appear to Western eyes as a form of dualism and are quickly dismissed. . . . It is important to note that these levels of subtle mind are not theoretical; instead, they are delineated rather precisely on the basis of actual experience, and they merit respectful attention by anybody who claims to rely on empirical science. . . . An understanding of these levels of subtle mind requires a sustained, disciplined, and well-informed meditation practice. In a sense, these phenomena are open only to those who are willing to carry out the experiments, as it were. That some form of special training is needed for firsthand experience of new realms of phenomena is not surprising. . . . But in traditional science, such phenomena remain hidden from view, since most scientists still avoid any disciplined*

study of their own experience, whether through meditation or other introspective methods. Fortunately, contemporary discourse on the science of consciousness increasingly relies on experiential evidence, and some scientists are beginning to be more flexible in their attitude toward the firsthand investigation of consciousness.[16]

11 ROBOTS THAT THINK THEY CAN THINK?

Does the brain work like a computer? Could the ability to invest the world around us with meaning emerge in robots equipped with artificial intelligence? Or is the quest to create artificial intelligence of this sort fundamentally misguided? What is the purpose, if there is one, of reflexive human consciousness, which allows us to question the meaning of life and our place in the world?

THUAN: As we've seen, some biologists think that organization of the matter in the brain only needs to become sufficiently complex in order for consciousness to emerge, along with thoughts and all the emotions, such as love, that make life worth living. In their opinion, there's no reason why consciousness wouldn't simply appear once evolution passed a certain threshold of complexity. The brain, in this view, is a thinking machine, the sum of its neuron parts. The relationships among neurons create what we call the "mind."[1]

This computational model of consciousness compares the neuron system to a computer's hardware and the mind to its software. The networks of neurons are the material framework of consciousness, just as a computer's electronic circuits provide the framework for the software that controls the machine. This purely reductionist explanation comes down to saying that if machines become sufficiently complex, then one day they'll be able to think and feel.[2] Some researchers into artificial intelligence are confident that one day they will be able to build robotic computers that will have feelings, and so will experience love and hatred, sorrow and pity. Nothing would then stop them from being creative, or from writing the next *War and Peace* or composing Beethoven's Ninth Symphony.

MATTHIEU: Even if a machine could mimic consciousness, that wouldn't make the slightest difference in its fundamental nature. All such a machine can do is process information that remains meaningless to it. Even if a machine were brilliantly programmed to "create" a new symphony, then it could do it only by following harmonic rules chosen according to the musical tastes of its programmers. Not only would the machine have absolutely no interest in the beauty of the music, it wouldn't even know that it was music.

T: In 1997, world chess champion Garry Kasparov was beaten in a match by a supercomputer called Deep Blue. Some journalists interpreted the defeat as a blow to humanity. In fact, Deep Blue beat Kasparov only because it could analyze two hundred million positions per second, which allowed it to compare all of the possible outcomes over at least the next ten moves. A human player can anticipate only a few combinations, and uses experience and intuition to make the decision. So it was quite simply Deep Blue's extraordinary computing power that defeated Kasparov. Deep Blue was no more aware that it was playing chess than a plane knows that it's flying to New York, and it couldn't have cared less about winning or losing. It just blindly followed the instructions that a team of computer scientists had programmed into its electronic circuits.

The will to win, anxiety, nerves, tension, the regret after making a bad move, or the pleasure of coming up with a winning combination are all totally alien to Deep Blue. Maybe it's because Kasparov experienced these human emotions that he lost the match.

M: Even a simple pocket calculator can beat us handily when it comes to multiplying three-digit numbers! But all this has nothing to do with consciousness. We worry about computers, but they don't worry about us, and there's little chance that a computer will wonder one day if humans are conscious or not!

T: So long as computers remain merely complicated circuits with electronic currents obeying programs, they'll remain machines and won't be able to think, feel, love, or hate. They'll just go on blindly manipulating sequences of ones and zeros. The computer in fact is just a highly sophisticated version of the ancient abacus, which is still used by the Chinese and the Iranians. In this instrument, the ones are represented by beads that slide along metal bars, while the zeros are empty spaces. Instead of electronic components, the fingers of the hand move the beads, leaving spaces according to a strict set of rules. A computer calculates much more quickly than an abacus, but this doesn't make it any the more conscious.

The computational model of consciousness has been criticized by some researchers, such as Francisco Varela, because it doesn't sufficiently account for the role of the brain's interactions with the external world. In Varela's opinion, these interactions play an essential role, and the mind is created by a process of give-and-take between the brain and the environment that he calls "enaction." This give-and-take process allows the "meaning of the world" to emerge. As he says, "The brain exists in a body, the body exists in the world, and the organism moves, acts, reproduces itself, dreams, and imagines. It is from this permanent activity that the meaning of its world and of things emerges."[3]

M: Yes, these days most neurobiologists reject the simplicity of the computational model. They argue that the brain, whose learning capacity is practically unlimited, works not according to a simple binary language like that of ones and zeros in computers, but in a far more complex and interactive way. According to the "dynamic" model they've put forward, the interdependence and interaction of the brain's neuron networks create states of activity in the brain that can be identified as consciousness. They say that consciousness is an emergent property of the brain just as liquidity is an emergent property of a collection of water molecules. According to Francisco Varela, "It is on the basis of dynamic patterns of relations that one can see the inseparable relation between a mental state and the pattern underlying it (which involves also the organism in its whole involvement with the world). So whether these patterns are objective or not is a question to epistemology: If you are a substantialist, then nothing is objective unless material. I and many other scientists think that patterns of relations in nature are objective and yet totally nonmaterial, without the shadow of a problem."[4]

T: I agree that the analogy between a brain and a computer is extremely superficial. If we look deeper, we find that a brain's processing is nothing like a computer's. While the computer stores information in the form of binary sequences of ones and zeros, in the brain, no one has ever shown that neurons work in this binary way, storing information using open positions (corresponding to the number one) and closed positions (corresponding to zero). There are other important differences. The brain is self-programmed, while the computer is not. Whereas a computer has an autonomous memory, with independent "inputs" and "outputs," in the brain, the memory zone is the same as the thought zone. Also, once the wiring of a computer has been set up, it doesn't change. If a lone wire snaps or a single transistor stops working, it breaks down. But the brain comprises networks of neurons that can regenerate themselves and is extraordinarily adaptable. The brain evolves constantly during our lifetime, and does so extremely rapidly during our childhood. Cells die, and others are born. The brain lets the connections it doesn't use die. There is a sort of natural selection among the neurons.

The speed of information processing is also extremely different. In the brain, impulses travel at speeds up to a hundred meters per second, but in a computer, information travels much faster, at several thousand kilometers per second. This explains why a computer can perform certain tasks far more quickly than we can—for example, when manipulating figures. In contrast, a human brain is far better at exercises of synthesis—for example, the recognition of a face.

Of course, if we wanted to defend the "mind as a machine" view, we could argue that while computers aren't conscious yet, this is only because we don't yet know how to make them as complex as the brain's neuron system. After all, the human brain is the fruit of over a billion years of evolution, while the first computers were invented only in the 1950s.

At that time, the English mathematician Alan Turing suggested a simple test to gauge a machine's intelligence.[5] Let's suppose, he said, that we're conversing with two hidden partners. One of them is a human and the other a computer. If, during the course of the conversation, we're unable to distinguish which is which, then we'll have to conclude that the computer is as intelligent as a human. In 1980, however, the American philosopher John Searle took issue with Turing's test. He did so by proposing the following imaginary experiment, known as the "Chinese room" experiment.[6] Instead of a dialogue between a computer and a person, as in Turing's test, the dialogue is now between two persons. The experiment goes like this: I go and sit in a room where someone passes me questions written in Chinese through an opening in the wall. I have to answer, even though I don't understand a word of Chinese. To do so, I have a list of ready-made answers and instructions that allow me to match an answer to each question. I pass the answer back through the opening to the questioner, who does understand Chinese. In this way we can have a long chat. But I can't claim to understand Chinese, or to have thought out my answers as I would have done if I spoke the language. I just followed some instructions, just as a computer mechanically follows its program. The conclusion is that a computer doesn't think, even if it's correctly programmed and can provide the same answers that I do. Even though the debate between Searle and the defenders of Turing's test is far from over, I personally think that the philosopher's argument is convincing.

Turing predicted that by the year 2000, computers would be able to fool a questioner for about five minutes of dialogue. He was overly optimistic. We're still far from being capable of producing computers that can hold a conversation like a human being, particularly when they're asked to think about themselves.

M: If you ask a human a question that's puzzling, strange, or casts doubt on his fundamental principles or his way of seeing the world, he won't come back with some absurdly irrelevant answer. But that's what computers usually do when they don't find the right answer in their program. The conscious being will think it over, and so find a new way of looking at life. For a computer, the word "life" has no more meaning than the dictionary definition in its memory.

T: Computers can read writing, understand spoken instructions, make approximate translations from one language to another, and solve problems that have defied mathematicians for generations. However, they still have only very limited "senses." Computers can't "see" very well, and have problems recognizing the person talking to them. They can understand just a few thousand words—and then only if you speak slowly and clearly. And they reply in a distinctly metallic voice.

M: Those are just technical problems that may be solved one day. But there are far more important qualitative differences. If consciousness is reduced to the working of neurons, and the working of neurons to the properties of their atoms, there's no fundamental difference between a flesh-and-blood computer and a metal one. But a metal computer is no more conscious that it exists than is a bag of nails. Why is it that we humans wonder, "What is the ultimate nature of consciousness? Who am I? What's the meaning of life? What will happen to me after I die?" An artificial intelligence system has no reason to wonder about its nature and to spend hours investigating the fundamental nature of consciousness, as contemplative people do. A computer doesn't wonder what will happen after it's been unplugged. Some artificial intelligence systems can learn, but why should they worry about what will happen to them in the future, or be pleased about how well they're working right now?

T: Some artificial intelligence researchers argue that the ability to have such thoughts should be able to emerge from robotic systems, if they can learn from their environment, in a similar way that these mental abilities emerged during the evolution of living beings. For example, Rodney Brooks of MIT and his team think that if we make a machine that knows nothing about its environment, but is equipped with a powerful sensorimotor loop that allows it to gather information from the outside world, then it will rush around everywhere like an ant, exploring its environment, going from one room to the next, then round the garden, between the trees, avoiding holes. Gradually it will develop what is called a reaction/action loop that will become so effective that it'll be able to cope in any environment. In other words, even though the machine has no initial representation of the world, one will emerge as it works.[7]

The brain probably emerged in living beings precisely to generate just such a reaction/action capability. For example, the neuron system hasn't appeared in plants, fungi, and bacteria. But it has developed in animals because, in order to eat, they need to pursue their prey.[8] So they needed a brain system to connect their sensory organs to their muscles. Abstraction and reflection arrived much later. The neuron system took more than one and a half billion years to evolve. During the first three-quarters of this process, animals could only perform the most basic survival functions, such as running, hunting, and feeding. It was only about a million years ago that language, symbolic intelligence, and social interaction appeared in primates. In Francisco Varela's view, "The appearance of the mind was not a major leap, but the necessary continuation of incarnation in evolution."[9]

Artificial intelligence (AI) researchers have pursued two basic approaches in their quest to create thinking machines. In the first, they attempt to simulate the process of evolutionary selection. They build thousands of tiny robot-creatures that are just slightly different one from the other. They then release these robots and allow them to compete with one another in order to select the fittest ones, creating a kind of machine evolution. The second method, which Brooks among others has adopted, is to try to program into robots brainlike capacities such as memory, recognition of faces, or the ability to interact socially.

M: But even if these robots can "adapt" as planned, this still doesn't mean that they will develop consciousness. In fact, there really is no good scientific definition of consciousness, and no scientific means of detecting its presence or absence in anything whatsoever. Reductionists who see the mind as a machine don't have any explanation for *why* consciousness emerges either from the brain or from the brain's interaction with the environment. Those who argue for the emergence view don't know specifically what type of complexity is needed to produce consciousness. Both theories are just conjectures, because the nature of consciousness is such that you can't simply study it from the outside.

Like most philosophers of mind committed to scientific materialism, the American philosopher Daniel Dennett concedes that "with consciousness . . . we are still in a terrible muddle. Consciousness stands alone today as a topic that often leaves even the most sophisticated thinkers tongue-tied and confused."[10]

The fact that some people insist on considering consciousness from a "third-person" perspective is no surprise. Such resistance is probably due to feeling insecure at the prospect of having to let the mind itself deal with the mind through sustained contemplative training. This is the attitude of someone who would like to learn how to swim without ever getting wet. In the case of robots, as the American philosopher of science Alan Wallace says, "since modern science is so ignorant about the origins, nature, potentials, and causal efficacy of consciousness, the hypothesis that a robot might answer correctly all possible questions regarding consciousness is to attribute a godlike knowledge to a robot that its designers don't even remotely have."[11]

T: Perhaps AI researchers have been so ambitious because of the remarkable success they've had lately in approximating a certain degree of functioning that looks a great deal like primitive consciousness.

Recent work[12] on what's called "new artificial intelligence" has shown that a group of small robots can organize themselves by interacting and behaving in a way that seems to imply they're conscious. For example, they can decide as a group to accept new robots that are introduced by evaluating whether or not the newcomers will help them function better. They can also decide to reject those that they determine will not help. Today's most advanced robots probably have the level of consciousness of an insect, and great strides are being made to get close to that of a dog. But we should distinguish here between this kind of consciousness, which is called primary, and the kind of reflective consciousness the human mind is capable of. The selection process mentioned above can be compared with what we call instinctive actions, such as with Pavlov's dogs. We're still far away from creating elaborate languages and reflexive thought.

In our daily lives, 90 percent of our thought processes involve experience-based primary consciousness. Walking, taking a bus, or preparing a meal doesn't require reflexive thought. "Reflexive," in this sense, means the ability to examine oneself. Reflexive consciousness examines its own existence and wonders about its destiny. It seems to have appeared about a hundred thousand years ago (at the period of Cro-Magnon man) when people started to bury their dead. Anthropologists believe that the earliest signs of reflexive consciousness were the ability to imagine a world after death and to prepare the journey with rituals, the remains of which have been discovered in some of the earliest known graves. Consciousness of this kind is also expressed in the creation of art, such as the early cave paintings. The paleolithic artworks produced about forty thousand years ago in the caves of Chauvet and Lascaux in the south of France are considered to be some of the earliest artifacts of human consciousness.

These creative abilities and higher thought processes seem to be linked to language capacity, and it's this close connection with language that makes reflexive consciousness uniquely human. We are the only species to have elaborate languages. "If I had only my experiences, I would be more like a gorilla," to quote Varela again.[13] A gorilla can't examine itself, because its linguistic capacities are extremely basic. As far as reflexive consciousness is concerned, either you have it or you don't. It seems to be a matter of all or nothing. With robots, too, we're still far away from creating elaborate languages and reflexive thought.

Human consciousness guides our thoughts not only about ourselves but also about others, and about our environment and the passing of time. It also teaches us that each person is unique and irreplaceable, and that his or her death is an incurable tragedy.

So, the question comes down to—will robots be reflexively conscious?

M: Even if we found a way to create a "ghost in the machine" and consciousness could be embodied in a robot, the robot (whether it's made of circuit boards or neurons) wouldn't be the prime cause of that consciousness, just as the neuron circuitry of the human brain is not the prime cause of our consciousness. When the Dalai Lama was asked whether a robot could embody consciousness, he replied that it was conceivable, but he didn't see why a consciousness would choose to associate itself with a machine, or what sort of karma would take it there.

In addition to their failure to appreciate the true nature of consciousness, reductionist neurobiologists have failed to grapple with the issue of free will. The "man as machine" model claims that when we have the impression that we're thinking and deciding, we're really just perceiving the aftereffects of calculations made by the neuronal system. For example, a moment of doubt comes from the fact that the neuron system needs time to work out the best solution. When several circuits come together, we have the impression that we've made a decision, and we feel relieved. According to the neuroscientist David Potter: "One is led to wonder whether decisions are ever made in consciousness or whether the consciousness in which we take so much pride is simply a reporter function in the brain. Are decisions and emotions calculated by nerve cells whose behavior we cannot bring into consciousness and cannot control by conscious mechanisms? This is a very uncomfortable question in Western science."[14]

Some neurobiologists have actually come to the conclusion that free will is purely illusory. We have the *impression* that we're free and make decisions, they argue, because this sensation of being in command has had a favorable effect on how our species evolves.[15] It has given us a competitive advantage in the game of evolution. This comes down to thinking that we're the same as robots that mistake themselves for thinking creatures. This sort of explanation is inevitable in a model where consciousness is just a light that flashes on at the end of a chain of neurochemical reactions. We can even wonder why the light exists. If all the decisions are worked out by the neurons, what's the point of consciousness? According to this view, consciousness would have no real influence on the brain, but would be a mere passive witness, a powerless underling that imagines it's the emperor.

Yet, if I set out to prove that I have freedom of choice, I have no problem at all doing so. For example, I can put off indefinitely the moment of standing up from a chair, at least until I fall asleep or pass out. This same choice of restraint can be applied to all my impulses—thirst, hunger, bodily functions, and so on. The only point of this mental veto is to prove that I have freedom of choice. There is no other purpose for it, and in fact it goes against instinctive survival mechanisms, so it's preposterous to claim that this mental control has been produced by the brain's unconscious calculations. A lunatic might remain stuck on his chair, but a sane person would have no reason to do so except in order to prove that he had freedom of choice.

We can also ask ourselves where the idea of proving that consciousness exists comes from. How could something that doesn't exist want to prove that it exists? How could unconscious scientists have devised a science that makes them deny the existence of consciousness? Isn't there something logically wrong here? Is it even necessary to wonder if consciousness exists? Our first-person experience of life tells us that it does. What other world exists, apart from the one that we experience? Does the reality of a world that doesn't concern us at all have any meaning? Denying the existence of consciousness thus seems to be more a metaphysical choice than a scientific proof.

T: In any case, science doesn't yet know how we think and create, or how we experience feelings of love and hatred, beauty and ugliness, or joy and sadness. While this remains a mystery, it's difficult to deal with questions regarding the origin of consciousness.

M: The most hard-core proponents of scientific materialism, who call themselves eliminative materialists, declare that subjectively experienced mental states should be regarded as nonexistent, on the grounds that the descriptions of such states are irreducible to the language of neuroscience. But "in denying the very existence of mental states as they are experienced firsthand, eliminative materialists attempt to override experience on purely dogmatic grounds."[16]

One last point that provides powerful evidence for a nonmaterialist form of consciousness is the fact that a stream of consciousness is capable of undergoing a complete, instantaneous turnabout in the way it apprehends the world. Neither a computer nor a neuron system is absolutely malleable. It takes a lifetime to set up all the connections among the billions of neurons in the brain, through their own form of natural selection. As you've already said, some of them waste away, while others establish stable connections, thus favoring the best possible adaptation to external life, social relationships, survival of the species, personal happiness, and so on. This starts when the brain is formed in the fetus, and continues into adulthood. The brain is certainly remarkably flexible, as can be seen in the large-scale rearrangements of neuron connections that begin minutes after amputation of a finger or a leg.

However, it's hard to see how, at a given time in our existence, this sort of system can revolutionize the way we think and live in just a few instants.

For instance, there are cases of murderers who have lived for years in the grip of hatred, and have gone on killing in prison. Then, after a particular event or thought, they suddenly understand that their behavior is inhuman, and start imagining a totally different world. In a short period of time, they learn to live a contrasting life based on love and altruism. Such a turnabout should theoretically involve a large-scale reorganization of neuron connections. But, even though the brain is flexible, this can't happen at once. The subtle consciousness, however, being free of physical constraints, can easily change from one moment to the next.

T: Such a complete and instantaneous change in behavior can also be seen in people who are suddenly touched by religious faith. Some of them were previously utterly indifferent to metaphysical questions, but then suddenly experience a burning religious feeling that completely changes their way of life and thinking. This is the "grace" or "illumination" that has been so well described by the French poet Paul Claudel and American-French writer Julien Green.

M: Buddhism says that the experience we have of our own consciousness, and our ability to understand its basic nature through introspection, as well as to master it through contemplation, all indicate that there is a consciousness continuum that transcends the physical frameworks of the brain.

But we must always keep in mind that all of this realm of experience is part of relative truth. Mental events, discursive thought, hope and doubt, or the impulses and reasoning that lead us to make certain decisions are all part of ignorance and illusion. We become lost in the tides of thought, which we mistake for realities. Above such delusion, the only undeniable knowledge is pure awareness, which is free of concepts, conceits, and representations. The primordial simplicity of pure awareness needs no proof but itself. It is the highest point of direct experience, both indescribable and unimaginable. Whatever way you view the continuum of this pure awareness, it cannot be refuted, any more than nothingness can refute existence. Concepts are powerless before the ultimate nature of the mind. Its nature consumes them, as fire burns birds' feathers without leaving any ashes.

T: Given this belief in pure awareness, does Buddhism think that humans are at the summit of intelligent life, or, as I expect, are there more highly evolved beings?

M: Yes, there are, and Buddhas are one example of this. There's no reason to suppose that other worlds don't contain beings that are more highly evolved than we are. In our world, there are considerable differences of spiritual development between different individuals, over and above differences of intelligence. A Buddha's intellectual faculties, and his understanding of the nature of the mind and the mechanisms of happiness and suffering, are far sharper than in people who haven't purified their consciousness continuum.

T: So the vision and understanding of the world of a particular consciousness depend on that consciousness's degree of evolution? Are there levels of this evolution?

M: It's said that there are three worlds. There's the "world of desire," which includes humans. This world is so called because minds are the constant playthings of powerful emotions. Then there's the "world of forms," in which consciousness is subtler and less prone to emotional impulses. Finally, there's the "formless world," where consciousness isn't subject to a bodily form. But this sort of existence still belongs to the world that's conditioned by ignorance.

T: Can we speak of rebirths in the third case?

M: It's more appropriate to talk of successive states of existence. Consciousness remains in this formless state for some time before taking on another bodily form.

T: What determines how my consciousness will be reborn?

M: All through our lives, free will allows us to modify our stream of consciousness by means of our thoughts, as well as by the words and deeds that arise from our thoughts, and, in turn, condition them. We can either learn to see through the veils formed by hatred, pride, and greed, or else be blinded by them. These states are called "obscure," because they stop us from seeing the true nature of consciousness and of the things we perceive. They deprive us of our faculty of judgment and destroy our mind's natural serenity.

For a Buddhist, true spiritual transformation means a change in the stream of consciousness. Just as we can pollute a river's water by throwing waste into it, or else purify it by filtering, so we can make our consciousness continuum clearer or darker during our lives. If the continuum has been purified, then our next existence (or the next physical framework for our consciousness) will be an intelligent person able to carry on this process of transformation that has now begun. But if we have further darkened our continuum, we will next experience life as an animal, or in another state of limited intelligence in which there's practically no chance for us to transform ourselves.

T: So does Buddhism consider that animals are conscious? Is an earthworm or a mosquito conscious of its condition? Observation of certain animals shows that they experience feelings similar to ours. Anyone who's seen a bitch feeding its pups can have no doubt about its motherly love. Anyone who's heard the shrill squawks of a bird being chased by a cat can have no doubt about its panic. Anyone who's seen a dog leap up to welcome its master when he comes home can have no doubt about its joy and affection. It seems that some animals, and especially those that are nearest to us genetically, like chimpanzees (whose genome is 99.5 percent the same as ours), can create mental pictures and recognize abstract concepts such as shapes and colors. Some are even sensitive to beauty. Groups of chimpanzees have been observed sitting and gazing in wonder at a sunset. Specialists in animal behavior have found that there's no fundamental difference between certain psychic activities in animals such as dolphins, or primates, and us. So animals seem to have a primary consciousness, but, as opposed to mankind, it doesn't seem probable that they also have reflexive consciousness of themselves and of existence.

We're unlikely to see a chimp telling its life story or writing a book like Proust's *In Search of Lost Time.*

M: That's why Buddhism thinks that animals can't travel along the path of spiritual liberation. However, when they've eliminated the reasons that led them to exist as animals, then they can once more take advantage of the opportunities of a human existence. A man's intelligence can be used for destructive purposes, but it can also develop wide-ranging, impartial altruism, while an animal's cannot. The unique value of human existence is that it leads to suffering so great that we try to free ourselves from our condition, but not so crushing as to make it impossible to follow the spiritual path.

The ability to think about ourselves is the sign of reflective consciousness, but looking for happiness and fleeing suffering indicate a deeper aspect of consciousness. One of the Tibetan words for "animate beings" is *drowa* (literally "something that moves"). It refers to motion in a direction that's determined by a particular awareness. This motion can go from the simple tropism of an amoeba to a hermit's journey toward enlightenment, via the running of a deer or the work of our hands. Of course, there are exceptions, such as fixed animals (coral, mollusks, etc.), but in general this purposeful movement is one way to distinguish animals from plants (which Buddhism doesn't see as animate beings).

In this way, we can consider that animals have a basic consciousness, and that their desire to avoid suffering and be happy is just as legitimate as ours. This is one reason why using animals for our own purposes without any concern for their suffering and often at the cost of their lives is absolutely indefensible in ethical terms.

T: Is there a hierarchy of states of existence? Is the formless state higher?

M: Yes, but only relatively, given that it hasn't been freed of ignorance. So long as ignorance hasn't been eradicated and we're still attached to the ego and the world of phenomena, then we will always fall back into the sufferings of the conditioned world. The ultimate state we're looking for is perfect, totally unveiled knowledge.

T: Does this knowledge allow us to escape from the cycle of rebirths?

M: Someone who has reached enlightenment is purified of all the tendencies and karma that can lead to another rebirth in the vicious circle of the conditioned world. He's free not to be born again. But, through compassion, he seeks rebirth. So long as people suffer in the cycle of rebirths, the enlightened being will continue to reincarnate in order to guide them lovingly along the path of liberation.

T: Is this what's called a Bodhisattva?

M: Buddhism teaches that there are three attitudes that we can have toward beings: the attitude of a king, who establishes his own power before taking care of his subjects; the attitude of the ferryman, who reaches the farther shore at the same time as his passengers; and the attitude of the shepherd, who walks behind his flock and ensures that all his charges are safe before taking care of himself. The true Bodhisattva is like the shepherd. He is ready to renounce nirvana, or Buddhahood, in order to stay in samsara to help people. But this is just an image, illustrating altruistic courage. In fact, the Bodhisattva doesn't need to wait for all living creatures to be freed before reaching enlightenment himself. What's more, a perfect Buddha does far more good than a Bodhisattva does. He does it spontaneously, as the sun shines effortlessly. When we reach full enlightenment, we can't stop ourselves from feeling infinite compassion for all beings. It's also said that, as the moon is effortlessly reflected on all stretches of water, so the Buddha's compassion will be manifested in countless incarnations so long as living beings go on suffering. As Shantideva said in the Way of the Bodhisattva:

> *And now as long as space endures*
> *As long as there are beings to be found,*
> *May I continue likewise to remain*
> *To drive away the sorrows of the world.*[17]

T: One issue with this idea of rebirth is troubling to me. Given that world population is constantly growing, and that each stream of consciousness is associated with a particular being, does the universe contain an inexhaustible supply of streams of consciousness to feed this rampant increase? Has this supply existed for all time, since the beginning of the universe? Would that mean that there are numerous consciousnesses that haven't been associated with any body since the Big Bang?

M: One of physics' main postulates is that the total sum of the universe's mass and energy doesn't change. In the same way, if streams of consciousness are without beginning or end, there's no reason to suppose that new ones spring up out of nothingness. But this doesn't mean that their number is limited. The quantity of streams of consciousness in a given universe increases or decreases according to the availability of physical frameworks. If a stream of consciousness can't appear or disappear, it can, as I've already pointed out, transform itself either in a process of darkening that leads to suffering, or one of liberation that leads to enlightenment.

This process of transformation is one of the most difficult aspects of Buddhist thought for many to grasp. And yet we know that it's possible to change ourselves substantially, otherwise we wouldn't bother to study or train. We know that it's possible to pass from irritation to calm, from jealousy to friendliness, and from confusion to intellectual clarity. Buddhism knows that we can also go much farther than this and radically transform every aspect of our selves. But this can't be done without effort, by simply letting things be. That's why it's so important to master our minds, and not to allow ourselves to go on as usual in what we term "natural" or "spontaneous" behavior. True freedom is not the freedom to follow every thought that comes into our minds, but to take our lives in our own hands.

By examining the nature of thought and phenomena, it's possible gradually to reach what we call enlightenment—in other words, a clear and lucid understanding of the nature of things that admits no confusion. Even though it can't really be expressed in words, this enlightenment allowed the Buddha to point out the path he had taken to other "travelers." We call this path a science of the mind, which shows us how to distinguish between the workings of illusion and the workings of knowledge, and above all to put this understanding into practice. It's now up to us to make this journey a direct experience. As I've already said, the Buddha often insisted that no one should believe his teachings out of respect for him, but should check them first. Now, to the Buddha's mind, the consciousness continuum is a fact that he experienced and the result of deep knowledge, not a gratuitous intellectual construct.

T: Perhaps, but it's still rather difficult for an ordinary mortal to understand and accept. There's no scientific proof of these ideas, and we're far from being able to check experimentally the existence of a "stream of consciousness."

M: There are many things that ordinary mortals find hard to accept—such as most scientific results! Just take the notion of space-time, or quantum uncertainty! What matters isn't so much that everyone should be able to check a discovery's validity right away, but that everyone who has done the necessary work, no matter how hard and long it may be, comes to the same conclusion. In Buddhism we have three sorts of valid proof. The first is proof by direct experience—for example, when you see a fire, you're sure that it exists. The second is proof by inference—when you see smoke, you deduce that there's a fire, and if you go over to it, you find that you were right. The third is proof provided by reliable witnesses—it concerns points that we can't see for ourselves in our current state of knowledge. If the man in the street believes in electrons, it's because a large number of reliable scientists believe that they exist. He's sure that he'd reach the same conclusion, if he spent a few years learning physics. And if the man in the street now feels less sure about electrons being real entities, then it's because other equally reliable scientists have deduced from quantum physics that an electron is just an "observable" phenomenon, which can also appear as a wave.

In contemplative science, "reliable witness" covers the large number of contemplatives—and the Buddha most of all—who have come to the same conclusions after many years of inner transformation. These people also show remarkable qualities of rigor and integrity. It can happen—but fortunately it does so only rarely—that a scientist fakes his results in order to pretend that he's made an important discovery. But once other researchers have checked these results, the fraud is rejected by the entire scientific community. In the same way, the so-called "guru" who tries to fool people will take gullible individuals in for some time, but his spiritual inexperience and incoherent behavior will soon be shown up by comparing them with genuine contemplatives. It isn't hard to sort the grain from the chaff.

T: I'm curious to know more about what happens to the streams of consciousness that reach enlightenment. Do they go on coexisting with the other ones that haven't reached it yet?

M: Enlightenment means that the darkness of ignorance has been eradicated. The mind is then totally free of the veils that normally smother it. When a Buddha reaches enlightenment, he doesn't suddenly vanish from the universe as though by magic. On the contrary, the perfect knowledge that comes with enlightenment leads to a spontaneous feeling of endless compassion and expresses itself in teachings to help others follow the same path. When a Buddha leaves his body after death, his consciousness can remain on the level of the mind's ultimate nature. This is what we call the "absolute body." He has no reason to be reincarnated because of delusory mental factors (such as desire, hatred, or confusion). To return, however, to the image of the moon shining on water without leaving the heavens, a Buddha's enlightened consciousness will effortlessly appear in many forms so long as beings continue suffering in samsara.

T: This notion that our thoughts can plague us and cause us suffering reminds me of Freud. How does the Buddhist notion of the consciousness continuum compare to the Freudian concept of the subconscious? What does Buddhism think of Freud's notion of the repression of the libido or psychic energy, and his idea that the fundamental energy of life is sexuality.

M: Rather than the vocabulary of subconscious drives, we prefer the terms "tendency" and "impregnation." During our past existences, we acquired all sorts of habits that remain latent in our consciousness continuum and greatly influence how we think and act. Our most firmly anchored tendency is egocentricity—the belief in the existence of an ego that controls our world. Because sexual desire brings the five senses into play at once, it's certainly one of the most powerful of the many impulses of attraction and rejection that create this attachment to the ego. It's impossible to rid ourselves of these tendencies quickly just by intellectual thought. There are three general methods. The first is to cancel them out by cultivating opposing tendencies that act as antidotes. Altruism, tolerance, non-attachment, or reflection concerning the unpleasant aspects of the objects of our desires cure egoism, anger, and attachment. They also help to eliminate unconscious tendencies that lie in a sort of reservoir and are always ready to leap out. This is a long process, because our consciousness continuum has acquired many "folds" that have to be ironed out one by one.

The second way is to meditate on the emptiness of the inherent existence of our tendencies, impulses, and thoughts in general. The result of this meditation is a faster and fuller liberation, because it attacks the very root of those atavistic habits and thus allows us to eliminate them in one go. The third method—for those who are capable—consists in using those very tendencies themselves as catalysts in order to bring about a swift, complete transformation.

This third method is risky, something like trying to snatch a jewel from the top of a snake's head. The risk is that one may simply reinforce habitual and deluded thought patterns that are causing affliction. In that case, the whole point of this method, which is to free oneself from negative emotions, backfires.

T: Freud also proposed a theory on the role of dreams. What does Buddhism think about dreams and the role of sleep? Sleep is a basic human need. Without it, we die. Researchers have found that when we dream, our brains use that time to eliminate the toxins produced by its chemical activities. They have further determined that there are distinct stages of sleep, which are divided into periods of true sleep, during which the brain slows down, and false sleep (or REM—"rapid eye movement"), during which the brain operates in a way similar to its waking state and produces dreams. Some neurobiologists say that while dreaming, the brain assembles a collection of memories and mental images and forms them into a more or less coherent story.

Freud and Jung and their followers thought that dreams were manifestations of our unconscious needs, conflicts, and desires, and that they revealed much about our personality and character. Others hold that dreams are nothing more than reflex responses to sensory stimuli, or that they are the by-products of cerebral disturbances resulting from chemical or hormonal imbalances.

M: A detailed comparison between the Buddhist and scientific views of this subject was carried out during one of the "Mind and Life" encounters, in which the Dalai Lama met with groups of scientists. Transcripts of these meetings were published in book form under the title *Sleeping, Dreaming and Dying.*[18] Buddhism considers that there are four stages between waking and deep sleep, dreaming being the second one. We say that deep sleep is a rehearsal of death, whereas dreaming is a rehearsal of the intermediary stage between death and rebirth, or *bardo.* In this state, our mind casts up all sorts of images in the form of hallucinations that seem very real to us.

There are techniques that allow us to be conscious that we're dreaming, to transform the dream, and finally to create dreams as we wish, choosing the subject matter and scenario. Some meditators practice doing this for months. The aim is to see that all phenomena are like dreams and are illusory, and so we mustn't be as attached to them as we were before. The phenomenon of the lucid dream has also been studied by the cognitive sciences.[19] After a few weeks of training, subjects become able to signal with a small blink of their eye (a muscular activity that can be monitored within one's sleep) that they are aware that they are dreaming. Indeed, the recording of their brain waves confirms that they are in a dream state.

T: On a related note, how does Buddhism describe the meditative state? Does it give us a foretaste of enlightenment? Is a state in which the intellect is silenced and intuition takes over, in which the usual consciousness of time and space is surpassed and we become aware of the unity of ourselves and the world?

M: Buddhism and its contemplative science, of which meditation is a key component, are described as a path, because they lead to a gradual transformation of the way our minds work, and to a purification of our stream of consciousness, thus taking it from confusion to enlightenment. There are different stages during this journey, which, in brief terms, correspond to the stabilization of discursive thoughts, increasing lucidity and serenity, an ever more accurate vision of the nature of external phenomena and consciousness, and above all a gradual breaking away from the ways of thought that normally obscure our minds. At the end of the journey we discover the true nature of the mind, which is wisdom and compassion, emptiness and luminosity, freed of all mental fixations. Since there's no more need for an attachment to the ego, notions of "mine" and "yours" break down as well. It's obvious that only the mind can travel down this path and succeed in knowing itself.

12

THE GRAMMAR OF THE UNIVERSE

NATURAL LAWS, MATHEMATICS, AND THE WORLD OF IDEALS

Do natural laws govern the world, and if so, should we think that they are part of a higher order of reality, separate from the realm of everyday life? Are they part of a Platonic world of Ideals that the human mind can perceive only in rare moments of insight? Or does the human mind create these laws? If so, do they depend on our ways of perceiving the world and on our preconceptions? Does the Buddhist concept of interdependence offer a way out of these conundrums?

THUAN: The concept of a stream of consciousness that can exist apart from a body leads to another fascinating area of inquiry, about the types of reality that might exist apart from the material world. Western science is founded on the belief that there are natural laws that govern our universe. Most scientists think of these laws as abstract rules that shape our world and are not created by our minds, but are rather discovered by us. So we can inquire, where do these laws come from? Do they exist in some higher-order realm of reality that transcends our material world?

MATTHIEU: Perhaps we should first ask where this way of thinking about natural laws as immanent, inherently existing rules comes from. The belief in such laws is strongest in cultures that have been shaped by religions with a belief in a Creator who is said to have established the natural laws so that there would be order in the harmony of the universe. According to this way of thinking, these laws not only *describe* how nature behaves and allow us to understand its behavior; they actually *dictate* that behavior. When did this idea of scientific law first appear in the West?

T: Our ancestors were already aware of certain regularities in nature tens of thousands of years ago. The alignments of standing stones and dolmens in Brittany (a region of France) and at Stonehenge in England, to mark the rising and setting of the sun at certain times of year constitute just one example. But many natural phenomena seemed completely mysterious to them. Anthropologists think that early humans lived in what appeared to them to be a magical world, in which many objects were inhabited by spirits. The Sun Spirit shed light on the Earth Spirit and the Tree, Flower, and River Spirits during the day. Then the Moon Spirit illumined the night. This spirit world was a re-creation and a reflection of the human world. The inanimate world was personified.

As we learned more, we began to perceive how small we were by comparison to the vastness of the cosmos. The personified universe developed into a grander mythical universe governed by gods with superhuman powers. According to these myths, all natural phenomena, including the origin of the universe, were caused by these gods' loves and hatreds. For example, in Egyptian mythology, the first being, Atum, contained all of existence. He engendered the earth and some eight hundred gods and goddesses. The sky was represented by the body of the fair goddess Nut, decked with glittering jewels representing the stars and planets. The Sun God Ra journeyed over Nut's body during the day to return at night through the underworld waters.

Such myths were the source of religions, which began to develop because people came to believe that communication with the superhuman gods required the intervention of priviledged intermediates, the priests.

M: I wonder if it's really so easy to know how our ancestors thought, ten or twenty thousand years ago! Just look at all the distortions of contemporary people's thoughts that occur within our own time.

We must be careful to acknowledge that there were many different ways of thinking through time, which overlapped. The philosophers of ancient metaphysics thought in very different ways from the believers who participated in spirit cults. Religions themselves differed substantially. Some developed profound metaphysical symbolism in their rituals, while in others such symbolism either never developed or its transmission has been lost. Based on the often simplistic interpretations of the Tibetan pantheon I've encountered, I wonder whether the Egyptian mythology you describe wasn't in fact quite a bit more sophisticated than this simple description suggests. Mythological images can seem crudely naive, though they may well reflect much more complex ideas.

Take Buddhist cosmology, for example. We talk of a central mountain, Meru, which is supposed to be the axis of the world, surrounded by four continents, with the Sun and the Moon orbiting around. This description could of course be written off as being outdated. But we would then miss out on the inner meaning, which reveals that Mount Meru is our spinal cord and the four continents our limbs, the Sun and Moon our eyes, and so on. This interpretive level shows the connection between our body and the universe. It goes beyond a simple cosmogony to give us an object of meditation.

T: Yes, there's almost always a metaphysical ingredient in myths. Certainly the tales and legends characteristic of the early belief systems were a way for people to give a meaning to the universe and the human condition.

But a very different style of thinking about how to understand the world eventually developed. In about the sixth century B.C.E., the Greeks came up with a radically new notion: nature was governed by laws that human reason was capable of perceiving, and was no longer the exclusive province of gods.

Aristotle sought to understand the behavior of natural systems by looking for an ultimate reason, or purpose, for the way they worked. He developed a theory of causality that was premised on four basic types of causes: material, efficient, formal, and final. For example, he would answer the question "Why is it raining?" by distinguishing the material cause, which are the drops of rain, the efficient cause, which is the condensation of water vapor into raindrops, and the formal cause, which is that they fall to the ground. But instead of using gravity, as a modern physicist would, to explain why rain falls to the Earth, he resorted to a final cause: rain falls because plants and living creatures need water to live and grow. The notion of natural laws is also found in the writings of Epicurus, Lucretius, and Archimedes, among others.

Oddly enough, however, the idea of natural laws really only took hold over human thought with the rise of the great monotheistic religions, like Christianity and Islam. They introduced the idea that God is separate from his creation, and thus governs nature through divine decrees. Nature's laws were thus no longer inherent in physical systems, but were imposed by a supreme being.

When modern science emerged in Renaissance Europe during the sixteenth century, early researchers such as Kepler and later Newton were deeply convinced that natural order reflected a vast divine plan, and that they were exalting the glory of God by revealing it. So Western science was triggered by the idea that there are laws imposed by a Creator. Renaissance Europe was imbued with the idea that God manifested himself in the rationality of nature.

This perhaps explains why science didn't take off in the East—in China, for example, even though it has an ancient culture that is complex and sophisticated, and was technologically ahead of the West in many fields. Gunpowder and the compass, among other things, were invented there. In the Chinese notion of nature, the world wasn't created by a god who handed out laws. It came about from the mutual dynamic interaction of two opposite forces, yin and yang. Since the notion of natural laws was foreign to them, the Chinese had no incentive to look for any!

M: This lack of development of the methods of modern science may have less to do with an inability to analyze phenomena than with a different scale of priorities as regards the various fields of knowledge. Which is more important—to know the mass and charge of an electron and to study the details of the world around us, or to concentrate on developing the art of living, to deepen our knowledge of vital questions such as ethics, happiness, death . . . and to analyze the ultimate nature of reality?

In Tibet, those who chose to leave the world behind and live in monasteries in order to deepen their contemplative research didn't do so because they couldn't think of anything better to do. They were often society's most brilliant members. They weren't trying to "flee from the world," but rather to make the most of their tranquillity in order to devote each moment of their lives to developing the human and spiritual qualities that would allow them to be of more help to others later on. There are two reasons why people don't discover the laws of nature: either they can't, or else they're busy doing something else.

T: And yet Buddhism has spent considerable energies on examining the laws of nature. What does Buddhism say about this notion of natural laws?

M: Buddhism admits that we can use laws to develop and express our understanding of the apparent reality of our world. But these laws have no inherent existence. If they had, then this would imply that they could exist without the world of phenomena.[1]

In terms of conventional reality, Buddhism accepts everything that can be shown using logic and valid knowledge. By "valid knowledge" we mean what can be perceived directly or else deduced by inference, and what can be accepted on the basis of reliable reports. But we believe that the laws of this relative truth are neither arbitrary nor dictated by a supreme entity. Nor do we ascribe to teleological thinking, always looking for the purpose behind the way things are.

To the physical laws, Buddhism adds the laws of karma, in order to explain the consequences of our positive and negative actions in terms of happiness and suffering. We must remember, however, that all of these laws pertain to the world of relative truth.

On the level of absolute truth, Buddhism refuses to accept that things are what they seem, and it examines their ultimate way of being. This analysis has led it to the conclusion that the phenomena we perceive in the world are just appearances, whose properties and characteristics don't belong to them intrinsically. That's why, by using poetic images, the Buddha compared phenomena to dreams and illusions:

> *Like a flickering star, a mirage or a flame,*
> *Like a magical illusion, a dewdrop, or a bubble on a stream,*
> *Like a dream, a flash of lightning, or a cloud*
> *See all compounded things as being like these.*

T: I think another reason that Eastern thought developed such different methods of inquiry than those of the Western scientific method is the belief in a holistic vision of the world. According to this vision, each part of nature interacts with all of the others, thus forming a harmonious whole. This notion discouraged the idea that is fundamental to the reductionist scientific method, that nature could be split into parts and each part be studied separately. If it were impossible to understand a small part of the universe without also understanding everything, science wouldn't be able to progress. I'd be unable to function effectively as an astrophysicist if I had to study the gravitational interactions of all the stars and galaxies in the universe in order to understand why the Earth revolves round the Sun. To advance in my work, I don't need to solve all the problems of the cosmos in one go!

In fact, modern science has confirmed that the universe does have holistic properties and that it forms a closely connected whole. However, despite the universe's global interconnected nature, the reductionist method still works. We could imagine a universe in which each physical phenomenon in a given place is so tightly connected with the rest that it would be pointless to study it before we'd grasped the whole. It would then be impossible to formulate simple laws. Knowledge of the universe would then be all or nothing. Yet science has allowed us to grasp scraps of information without knowing the entire story, to hear a few notes of music without listening to the entire melody. The reductionist method allows us to progress step by step, putting together the jigsaw piece by piece, but without seeing the final picture.

M: There's a Buddhist proverb that says, "All large difficult tasks can be divided into small easy tasks." That said, the basic Buddhist notion of interdependence overrides this compartmentalized approach. The world isn't made up of a set of separate objects that have been put together into a large machine, like the gears of a clock. As Heisenberg wrote, "It must be stressed that the hope of gaining understanding of the whole world from a small part of it can never be supported rationally."[2]

The reductionist approach works very well if your aim is to gather information about observable phenomena, then piece them together so as to come up with laws that allow us to predict how they work, but without necessarily understanding their true nature. But if we examine the nature of reality itself, its ultimate existence, then looking at an atom or the entire universe comes down to much the same thing. Aryadeva wrote in the Four Hundred Verses, "He who sees the ultimate nature of one thing, sees the nature of all things." Drinking a drop of seawater allows us to deduce that all of the sea is salty.

The reductionist method is ultimately limiting. By so strongly emphasizing smaller parts of the complete picture, we become submerged under a mass of descriptive data and may forget to examine the very nature of things and ask questions about the "realistic" vision of the world.

T: Yes, by always peering at details, we run the risk of no longer seeing the whole. Excessive specialization is one of the tendencies that I most deplore in modern science. Some physicists can know everything in their tiny fields of study, yet be totally ignorant of other areas of physics. It often happens that articles written after years of research can be understood by just a handful of experts working in the same field. I'm nostalgic for the time when Leonardo da Vinci, Descartes, and Pascal were conversant with most of the knowledge available in their times.

But the fact is that reductionism has allowed huge progress to be made. This success stems largely from the fact that physicists of genius, such as Newton, have managed to isolate certain phenomena, or parts of the world, that do seem to follow a set of precise, linear laws.

We call a physical system linear when the whole is exactly equal to the sum of its parts—no more, no less. In systems of this sort, a given number of causes lead to a corresponding number of effects. So we can study the individual behavior of each part, then put them all together so as to deduce how the entire system behaves. For example, a rubber band lengthens a certain amount when you pull it using a given force. If you double the force, its length also doubles. We call this behavior linear because, in a graph, if you plot the length of the rubber band on the Y axis, and the force applied on the X asis, you obtain a straight line. It's that same linear property, but of sounds, that lets you hear the difference between a violin and a piano when you listen to a Tchaikovsky concerto: the sounds mingle, but don't lose their identities. In the same way, the linearity of light allows you, during the day, to see a weak red light that mixes with bright sunshine but isn't swamped by it.

M: But such linearity is simply a special case, and it doesn't apply to every phenomenon.

T: Quite. But reductionist physics was so successful that until the end of the nineteenth century it created the impression that all the world's systems were linear. Now we know that this is far from being the case. Nearly all systems become nonlinear past a certain limit. If you continue pulling a rubber band, eventually it no longer lengthens, it abruptly snaps. All chaotic phenomena are nonlinear. Our lives are full of nonlinear situations. For example, the brain certainly doesn't work in a linear way. Or, to return to the example of an orchestra, the pleasure resulting from listening to a Beethoven symphony exceeds the sum of the pleasures of listening to each instrument separately, and the pleasure of listening to a melody is greater than the sum total of hearing each note by itself. In some systems, when all the pieces are put together, "emergent" properties appear, so that the whole is more than the sum of its parts.

M: A community's power is greater than the sum of its members' various faculties. Buddhism illustrates this with a mundane yet eloquent image: you would try in vain to sweep up dust with a hundred separate twigs, but if you put them together into a broom, the job becomes easy.

The linearity of phenomena can thus be only an approximation—a theoretical case applied to a Platonic "Ideal" of your rubber band. In truth, as the rubber band is stretched and stretched, it is changed by this stretching. The rubber band also ages. It can't possibly always lengthen in direct proportion to the force used. In Buddhism, no system is really linear. That would imply an artificial world run by perfectly constant forces, wielded by immutable entities. But no permanent entities can exist in a stream of constantly changing, interdependent phenomena.

T: As far as physicists are concerned, the aging of a rubber band is imperceptible in the time their measurements last. And so describing a rubber band as a linear system is an excellent approximation.

The other key reason why reductionism has been so successful, despite the holistic nature of the world, is that so many phenomena are so heavily governed by what we call local forces. The behavior of many things we want to study is mainly determined by forces or influences in their immediate environment.

In fact, physicists argue that every natural phenomenon can be explained in terms of the four fundamental forces: the weak and strong nuclear forces, the electromagnetic force, and gravity. The strong nuclear force holds atomic nuclei together. It's the glue of protons and neutrons, as well as of the quarks that constitute them. Its range is tiny: it acts only at the scale of an atomic nucleus—or a ten-thousandth of a billionth of a centimeter. The weak nuclear force is responsible for radioactivity, in other words the transmutation of an atomic nucleus that spontaneously loses part of its mass when emitting particles or electromagnetic radiation. The range of the weak force is smaller still, only one-tenth that of the strong force. The electromagnetic force holds together the atoms and molecules, and the double helixes of DNA. Its range is on the level of everyday objects, and it determines the shape of a rose petal or the outline of a Rodin statue. It makes things solid and stops us from walking through walls or passing our hands through the pages of this book.

Finally, gravity pins us to the ground, stops us from floating in midair, and makes us fall to the ground when we trip. Its range covers the entire universe and is responsible for its overall architecture. It makes the planets revolve around the Sun, and connects the Sun to hundreds of billions of other stars to form our galaxy, the Milky Way. It groups together thousands of galaxies into clusters and tens of these clusters into superclusters.

The ranges of the electromagnetic and gravitational forces are, in principle, infinite. But the intensity of the electromagnetic force drops in inverse proportion to the square of the distance between two electric charges, and gravity by the inverse of the square of the distance between two masses. So their influence at a given place is limited to the local environment and does not extend to far-off objects, because the forces between two distant charges or masses are practically zero. For example, when an apple falls to the ground in an orchard, this is mainly due to Earth's gravitational pull. The influence of the Moon, Sun, or other celestial bodies is tiny and can be neglected.

M: Yes, but no object in the universe is governed *exclusively* by local influences. In the Buddhist view, this is in an important point, because if an object did exist that was totally independent from the rest of the universe, then it couldn't interact with the other parts. In a sense, *it would not belong to the universe,* which comes down to saying that it wouldn't exist. If it existed alone, then it would either be its own cause, or else causeless, which is absurd.

T: Of course, there are nonlocal influences, as we saw during our discussion of interdependence and the global nature of phenomena, the EPR phenomenon, Foucault's pendulum, and Mach's principle. There are mysterious immanent and omnipresent interactions in the universe, which require no force or exchange of energy, and which physics is at present incapable of describing.

Yet we find that the influences of faraway objects mediated by the four fundamental forces are for the most part so weak that we can eliminate them from our calculations. For example, to calculate the orbits of planetary probes, NASA needs only to consider the gravitational pull of the Sun and its retinue of nine planets. Gravity from the hundreds of billions of stars in the Milky Way can be disregarded. How can Buddhism gain knowledge of nature without resorting to the reductionist method? Does it think that we can't understand a part of nature without understanding the whole?

M: Buddhism doesn't reject reductionism as a tool for understanding nature's mechanisms. But its basic approach is quite different. It's far more concerned about how an erroneous idea about the existence of phenomena can influence our lives in terms of happiness or suffering. In this sense, it's more worried about a *reifying* approach than about reductionism itself. As we've already said, reification means giving an intrinsic existence to the qualities and characteristics that our common sense perceives. This process generally also leads to making consciousness into a concrete object, and to the conception of the mind/body divide. Almost all people engage in this kind of reification all the time in their ordinary perceptions of things, without realizing it. It also dominates science, even though it's no longer compatible with science's own recent developments. This disparity makes some scientists perform endless juggling acts while trying to reconcile the results of quantum mechanics and their solid vision of the macrocosm.

Buddhism favors instead the integration of the understanding of the unreality of things into our daily lives. The Middle Way, which Buddhism adopts by refuting both nihilism and realism, allows us to resolve many of the paradoxes that so puzzle scientists.

You mentioned earlier the notion of Platonic ideals. The Platonic conception of reality has been so compelling for so long that people steeped in this tradition of thought find the Buddhist conception of reality terribly difficult to accept. Central to Plato's concept of reality is the notion that there is a realm of pure truth, of ideals, that is separate from the human sphere. This is the alleged realm of natural laws; they act on but exist apart from our world. Buddhism has refuted in many ways this notion of separation and of ideals of any kind existing as immutable entities.

T: Plato and other philosophers supported their contention that there was such a separate world of perfection largely by pointing to the remarkable powers of mathematics to describe the concrete world. According to this view, mathematics is the language in which the natural laws are expressed.

Plato defined two levels of reality: first, the level of the physical world perceived by our senses, which can be measured and quantified, and which is impermanent, changing, ephemeral, and illusory; then the level of the true world of immutable and eternal Ideas. The temporal world we perceive is just a pale reflection of the world of Ideas. You know the famous allegory of the cave, which Plato used in his dialogue *The Republic* to show the dichotomy between these two worlds. Outside the cave, there is a vibrant world of colors, forms, and light, which men cannot see or reach. All they can observe are shadows projected by the objects and beings of the outside world on the walls of the cave. Instead of the brilliant colors and clear shapes of the glorious reality, all they have are the sullen darkness and the indistinct outlines of the shadows. Plato thought that, as in the shadow world, the universe we perceive is merely an impoverished version of the world of Ideas. Illuminated by the Sun of intelligence, this world of Ideas was also the domain of perfect mathematical relationships and geometrical structures.

The belief that the regularity of these underlying relationships can be described in mathematical terms constitutes the foundation of the scientific method. Some scientists even make the rather exaggerated claim that results that can't be expressed in mathematical language can't be considered truly scientific.

M: Indeed, Galileo wrote that anything that didn't involve the study of the measurable, quantifiable properties of material bodies (properties such as volume, weight, speed of movement) isn't scientific. But this position considerably reduces the range of science. We shouldn't forget that the word *science* comes from the Latin root *scire,* "to know." To reduce knowledge to what can be described in mathematical equations is absurd. It excludes vast ranges of our experience of life. For example, understanding that goodness warms our heart and hatred pains us is part of knowledge and so also part of science. This truth can be seen repeatedly, its mechanism can be analyzed and its causes understood. For example, let's start with the hypothesis that the brain is fundamentally peaceful, and that animosity and jealousy are just fleeting disturbances that conceal its true nature. We can test this hypothesis by a contemplative experiment. What part can mathematics play here? Anything that we learn by means of experimentation and that has been tested methodically and rigorously can be considered scientific. The notion of an "exact science" shouldn't be restricted to quantifiable facts, accurate to ten decimal places. A science is exact if it comes to correct conclusions about the nature of things.

T: I agree that the idea that all scientific knowledge should be expressible in mathematical terms is ridiculous. I must admit, however, that the extraordinary success demonstrated by mathematics in describing reality is extremely mysterious, and that this seems a compelling reason to accept the idea that there is an abstract level or realm of truth, or of natural laws, that we humans can only see part of.

The ancient Greeks argued that the physical world was just the reflection of this mathematical order. "Number is the principle and source of all things," Pythagoras proclaimed in the sixth century B.C.E. Twenty-two centuries later, Galileo echoed this: "The Book of Nature is written in mathematical language." In the twentieth century, the physicist Eugene Wigner expressed his amazement at the "unreasonable effectiveness of mathematics" in describing reality.[3]

The history of science abounds with examples of the extraordinary prescience of mathematics. In practically every instance where a new discovery has led physicists into uncharted waters, they have found that mathematicians have beaten them to it. For example, when in the 1910s Einstein discovered that gravity bent space, he couldn't use Euclidean geometry, which describes only flat space. So he was delighted when he came across the work of the mathematician Bernhard Riemann, who had developed the theory of curved geometries in the nineteenth century.[4]

In the 1970s, the mathematician Benoît Mandelbrot was looking for a new way to describe the geometry of irregular objects. Euclidean geometry works very well with straight lines, cubes, and spheres, but it becomes useless when studying shapes that are irregular, twisted, broken, discontinuous, or rugged. Yet most shapes in the real world are not regular. While Euclidean concepts, such as the straight line and the circle, are potent abstractions that have helped us make great strides in studying nature, they have their limitations. "Clouds aren't spherical, mountains aren't cones, and lightning doesn't come in straight lines," as Mandelbrot likes to point out. So, in order to describe the geometry of the irregular, he introduced the concept of "fractional dimensions"—the dimensions of an irregular object are no longer represented by whole numbers such as one, two, and three, but by fractions. These are "fractal" objects. There, too, Mandelbrot found out that the idea of fractional dimensions had already been suggested by the mathematician Felix Hausdorff in 1919.[5]

Why are abstract concepts, which spring out of mathematicians' minds, and are generally useless in everyday terms, so often in agreement with natural phenomena? So much so that physicists are taken aback when a new physical theory appears that does not yet have the mathematical concepts to go along with it, such as superstring theory. One explanation for this uncanny agreement is that originated by Pythagoras, that all the world is really a reflection of the world of mathematics.

So we come to the question, What is the nature of mathematics exactly? Within the community of scholars of math, there are two contrasting views. In the opinion of the *constructivists,* math doesn't really exist. According to the philosopher David Hume, "All our ideas are merely copies of our impressions." The only reality of geometrical forms is in the forms of nature. On the opposing side we have the *realists,* who think that math has a "reality" that is distinct from our thought. It makes up a vast field, which we can explore and discover with our reason, just as an explorer can discover the Amazon Rain Forest. Mathematics exists, whether we are conscious of it or not. Many of the greatest mathematicians have shared this view. Listen to what Descartes had to say about geometrical figures: "When I imagine a triangle, even though there may not be, nor ever have been, one anywhere in the world except in my thought, this figure nevertheless has a certain nature and form, or determined essence, which is eternal and immutable, which I did not invent and which in no manner depends on my mind."[6]

M: If he didn't invent it, then I wonder who did. And if it doesn't depend on his mind, then he'd have problems thinking about it!

T: Nearer our own time, the English mathematician Roger Penrose has written: "There often does appear to be some profound reality about these mathematical concepts, going quite beyond the mental deliberations of any particular mathematician. It is as though human thought is, instead, being guided toward some external truth—a truth that has a reality of its own and that is revealed only partially to any one of us."[7]

This feeling of a mathematical reality that is independent from our minds becomes even stronger when we see how it lives independently from its creators and seems to be dragging them ineluctably toward the truth. The German physicist Heinrich Hertz put it this way: "We cannot help but think that mathematical formulae have a life of their own, that they know more than their discoverers do and that they return more to us than we have invested in them."

M: But I don't see at all why are you so surprised about the compatibility between concrete reality and mathematical abstraction. There's nothing odd about the fact that what we *conceive* corresponds to what we *perceive.* The way we explore the world, then sort out our perceptions of it, necessarily agrees with our mathematical concepts, because both perception and conception are products of our minds. To see the physical world as the reflection of mathematical order seems to me to be looking at things the wrong way around.

Buddhism would say that mathematics simply consists of human concepts applied to natural order, and this order is itself a reflection of interdependence and the laws of causality of which consciousness is a part. The fact that mathematical propositions come *before* or *after* the discovery of their natural equivalents makes no difference. It doesn't give them a special status or a fundamentally different existence. Is it really surprising that arithmetic can be applied to the number of stones on a path, or the notion of fractional dimensions to fractal objects? Arithmetic and geometry have no "inherent" existence either in our minds or in the external world.

T: And yet I don't think that the "unreasonable effectiveness" of math in describing the world comes from the interaction between our consciousness and the exterior world. Most math is worked out by mathematicians in a process of pure thought, in a completely abstract way.

M: But, when studying mathematics, mathematicians quite simply study how their minds work and how our intellects gradually come up with a way to interpret phenomena. This interpretative system is naturally lying ready to be used on new phenomena as they crop up. After all, that's its raison d'être. So it isn't surprising that this tool is sometimes ahead of our observations and can be used for purposes we never dreamed of. Mathematics can consider all sorts of logical possibilities that haven't found an equivalent phenomenon in nature yet, and may never do so. But that doesn't give these ideas an autonomous existence.

So far as Buddhism is concerned, "pure thought" isn't mathematical intelligence, or any other kind, but pure awareness, which is the mind's fundamental ability to be conscious. This "luminous" aspect of the mind differentiates it from the absence of consciousness in a stone, for instance. Our intelligence is conditioned by its physical framework—our bodies—and by the experience that our consciousness has acquired, though not only in this life. When physicists are surprised to find ready-made mathematical tools, it's like the surprise we feel when we discover that, unbeknownst to us, two people we know well are closely related.

T: In other words, given that our minds coexist with the world of phenomena, everything they conceive of must be in conformity with that world.

M: I don't mean that everything we dream up has an equivalent in the universe! But any coherent conceptual system, be it logical or mathematical, must necessarily reflect the interaction between our consciousness and the world, for the simple reason that we can't separate them without destroying them.

T: Personally, I still prefer to turn to Plato and the notion of a world of pure mathematical ideals.

M: Buddhism's position is clear on this point. Although it never debated with Greek philosophers, it did come up against Hindu thinkers with very similar ideas. Buddhism logically refuted the notion of an Idea. For example, some people in India thought that the Idea "tree" was the essential principle of all existing trees, which were thus simply "special cases" or gross manifestations of that Idea.

T: That's Plato's argument, practically word for word.

M: If the "tree" principle didn't exist, the Hindu philosophers claimed, we couldn't conceive of the abstract notion of a tree. It is at once independent of any particular tree and applicable to all of them. The Buddhists replied that we must choose: either this Idea "tree" has an existential link with natural trees, or it doesn't. If it does, then this link must appear in some phenomena that we can experience. But that isn't the case. If the Idea "tree" were intrinsically connected to all trees, then they should all grow when one tree grows, and all die when one tree dies.

If this idea has no connection with natural trees, then it serves no purpose. It's as imaginary as a tortoise's fur or a hare's horns. So we can get by without it. What's more, how could an *immutable* Idea interface with a *transient* object? Ideas are thus mental labels, nothing more.

T: Plato says quite the opposite. He makes a distinction between the perfect Idea of a tree and all its imperfect manifestations in the world. He postulates two god figures. One is the Good, eternal and immutable, who rules the world of Ideas. The other is the Demiurge, who fashions the matter of the contingent and changing world—of trees, for example—in conformity with the perfect forms of the world of Ideas.

M: Is there a connection between the Idea of a tree and its manifestation? If there's none, then the Idea is useless. Why, then, consider that it exists?

T: Plato would have replied that it defines general and ideal characteristics.

M: This is a mere mental construct, no matter how useful it might be in terms of classifying certain plants as members of the "tree" family. But the existence or nonexistence of this Idea has no influence on trees. And if this Idea *does* have an interface with gross reality, it can't be immutable.

T: You've put your finger on the heart of the problem. There's a profound dichotomy between the changing world of experiences and the immutable world of Ideas. Plato didn't try to reconcile them, but simply postulated two god figures, while declaring that only the Good was true and the Demiurge a mere pale and illusory copy. Christianity tried to solve this problem by bringing in a God who is outside time and space and who creates the world *ex nihilo*.

The discoveries of modern science suggest a different solution. Nature is governed by laws of organization and complexity that are outside time, and are thus immutable and unvarying as well. But, despite this, the world isn't immutable. It can change because, thanks to quantum uncertainty and chaos, the universe can freely express its creativity by improvising around those unchanging laws. By choosing among a large range of possibilities, the universe can thus be both changing and contingent.

M: An invariant that has "designs"? Designs imply plans for the future, so there goes the notion of timelessness. Also, how could this invariant then be unchanging? Laws that are outside time? Why not "outside phenomena"? And in that case, what do they apply to? Something that is immutable and outside time remains that way, it doesn't create a universe.

T: I must admit that the scientific concept of immutable laws doesn't solve the dilemma of how an immutable God could create a changing universe, and yet still be a part of that universe. Perhaps we can conjecture that God has no need to remain within time. He can turn away from his creation and has no more need to participate in it.

M: But by turning away from his creation, God loses his omnipotence. From a Buddhist point of view, all this makes little sense. An entity that is immutable is alone, remains alone, and has nothing to do with our universe. Why must we always insist on freezing the exuberance of true magic—the unreal unfolding of an infinity of phenomena—into such concepts? Why should abstractions exist ontologically? Isn't being a mental construct the very nature of abstraction?

T: And yet, how do we explain the fact that mathematical intuitions sometimes spring up, suddenly and unexpectedly, in a completely spontaneous way and without any preparation? This sudden contact with the world of mathematical concepts can take place at the most unexpected moments, as with Archimedes crying "Eureka!" in his bathtub. Henri Poincaré told how the solution to a mathematics problem, which had eluded him for weeks, suddenly came as clear as day into his mind, with no apparent preparation, when he was least expecting it:

> *Just at this time, I left Caen, where I was living, to go on a geologic excursion under the auspices of the School of Mines. The incidents of the travel made me forget my mathematical work. Having reached Coutances, we entered an omnibus to go some place or other. At the moment when I put my foot on the step, the idea came to me, without anything in my former thoughts seeming to have paved the way for it. I did not verify the idea; I should not have had time, as, upon taking my seat in the omnibus, I went on with a conversation already commenced, but I felt perfect certainty. On my return to Caen, for conscience' sake, I verified the result at my leisure.*[8]

Suddenness, brevity, and immediate certitude are the characteristics of mathematical intuition. To me, this spontaneous insight supports the idea that when the mind makes mathematical discoveries, it enters into contact with a realm of Platonic mathematical concepts. Roger Penrose is adamant about this:

> *I imagine that whenever a mind perceives a mathematical idea, it makes contact with Plato's world of mathematical concepts. . . . When mathematicians communicate, this is made possible by each one having a* direct route to truth . . . *because each is directly in contact with the same externally existing Platonic world! All the information was there all the time. It was just a matter of putting things together and "seeing" the answer. . . . In their greatest works, mathematicians are revealing eternal truths that have some kind of prior ethereal existence.*[9]

M: Frankly, I think intuition can be explained far more simply. After all, if we concede that mathematical insights are pulled from this world of ideals, then we would have to conceive of a vast range of such ideal worlds of insights. We'd have to envisage a world of Poetic Ideas, which has been visited by Baudelaire, Tagore, Rilke, and so many other great poets; then a world of Decision Ideas for undecided people who suddenly make up their minds, and so on. Plato's Ideals are reflections of the belief in prime, immutable causes that work only in one direction.

The fact that mathematics is so effective in describing the world, and that certain people are exceptionally talented at perceiving mathematical truths, simply shows that math is a part of our world, that it depends both on the world and on our consciousness.

The consciousness that conceives mathematics isn't exterior to nature. The way we perceive the world is closely linked to the workings of our minds—to such an extent that some schools of Buddhist thought have even called the external world an "image of our thought." Of course, some neurobiologists claim that the opposite is true, and that our mental constructs are "traces" left by the exterior world on our neuronal system. In fact, interdependence shows that the influence is mutual, and that mathematics is just one reflection among others of interdependence.

Interdependence transcends the division between the "exterior" and the "interior." The mathematical intuitions that some people have simply reflect the natural interpenetration of consciousness and the world of phenomena. In fact, what we should be wondering about is the origin of the illusory splitting of this complete interpenetration. In Buddhism, the dichotomy between "me" and "the world" is the first sign of ignorance. In a sense, it's Buddhism's "original sin," but it's original only in name, given that we commit it every moment of our lives.

T: How, then, does Buddhism explain there aren't more who are able to discover mathematical truths? Not all people have seen what such great mathematicians as Riemann or Srinivasa Ramanujan have perceived. Ramanujan's career in particular is an astonishing testament to the spontaneous quality of mathematical intuition. He was born into a poor family in Madras, India, where he received only a very rudimentary education. He was basically self-taught, and he rediscovered in his own way, and in the greatest intellectual isolation, a large number of well-known mathematical results. What's more, in a totally intuitive manner, and with no rigorous demonstration, he formulated hundreds of theorems that still today, more than fifty years after his death, continue to puzzle us. The problems that Ramanujan examined in such an original and intuitive way were often the very same ones being investigated by traditional mathematicians at the time.

So here we have a man with a radically different social and cultural background, and with no academic education, who nevertheless still finds the same mathematical ideas as his highly educated, more-conventional counterparts. I can't help thinking that he drew his inspiration from the same Platonic world of mathematical concepts as his colleagues. In the same way, when I hear of people capable of feats of mental arithmetic, or "autistic prodigies" who, despite their handicap, manage to solve extremely complex mathematical problems, I think of that access to the world of Ideas.

M: I think that it's rather like someone seeing himself in the mirror and not recognizing himself. Mathematicians communicate with the mathematical ideas in their head. Biology has recently discovered that people who are good at mathematics and mental arithmetic have differences in the regions of the brain corresponding to vision and language. This seems to allow them almost literally to visualize mathematical relationships that normally elude us. Einstein said that he sometimes had the impression that he "saw" the answer to a problem. There are other cases, just as startling as Ramanujan's—for example, a pair of twins who, after five minutes' thought, were able to list all the prime numbers that comprise twenty-five digits.[10] The psychologist who was studying them—I should say in passing that their IQ was quite low—says that one day he spilled a box of matches over the floor. The twins exclaimed, in unison, "111!"—the exact number of matches, which they could see as clearly as we can see that there are four glasses on the table.

So biologists have put forward the idea that mathematics is closely linked to the workings of our brain.[11] This idea corresponds to that of Buddhism, which says that mathematics is just one way to interpret phenomena, and that math does not derive from an inherently existing ideal truth.

According to our level of intelligence, mathematical concepts reveal certain aspects of the interdependence of phenomena. Poets interpret the correspondences between our minds and phenomena in terms of beauty.

A physicist expresses them in a mathematical formula. The universe isn't "too complex" for our understanding, because it's *our* understanding that determines *our* universe. The degree of complexity of natural laws reflects the level of intelligence of the mathematicians that formulated them. So it would be wrong to say that mathematics and natural laws are "smarter" than the people who conceived them. For someone untrained in mathematics, the formulas in physics don't describe the universe—they're meaningless. In the same way, laws aren't so complex that they elude the understanding of mathematicians, because they wouldn't then be able to formulate them.

There are large differences between people's cognitive faculties in various fields. Take contemplatives, for example. A beginner's intellectual understanding of emptiness is hardly comparable with an enlightened Buddha's knowledge, garnered from direct experience. We say that they are as different as the drawing of a lamp and the lamp itself. We can develop, train, purify, and transform our minds to a considerable extent. We can move gradually from total confusion, dominated by hatred and other mental poisons, to an intermediate stage of serenity, altruistic joy, and self-mastery, before finally arriving at enlightenment, which will give us a true vision of phenomena's ultimate nature. At that point, knowledge functions with an immediate certitude that transcends discursive thought.

T: To explain why some people are better at mathematics than others, you have referred to both neurobiology and to spiritual training. In neurobiology, the latest research into mathematical activity in the brain—which is still in its infancy—does indeed say that it results from a close collaboration between two areas in the brain: the one associated with vision (the two lower parietal lobes and the interparietal sulcus), which would thus be the motor of mathematical intuition; and the one devoted to language (the lower left frontal lobe), which would translate those intuitions into symbolic formulas. In 1999 a team of Canadian neurologists announced that those two parietal lobes in Einstein's brain were 15 percent larger than average, which perhaps explains his genius. For intuition plays a vital role in science, and great researchers have always made much use of it. Mathematical formulation (which involves the brain's language area) only comes in later to verify the intuition. In the story I mentioned earlier, Poincaré was immediately sure that the solution he'd glimpsed during his trip was right. He only checked it out when he got back home, using rigorous mathematical language, in order to have a clear conscience. Mathematical demonstration quite simply corroborated a result already provided by intuition. That said, neurobiology is still far from being able to describe exactly how we think or create.

M: In brief, for Buddhism the ability to perceive the harmony of the universe is an integral part of our minds. Formulating laws in terms of equations, numbers, relations, correspondences, and structures results from conceptual thought. But our conceptualization of what is in fact pure interdependence has no inherent existence.

13

REASON AND CONTEMPLATION

HOW WE LEARN ABOUT THE WORLD

What are the limitations of logic and discursive thought, which are foundations of the scientific method? Will science ever be able to answer all of our questions about our world and reveal an "ultimate truth"? In what ways does the rational, analytical approach of science differ from Buddhist contemplative science? How does Buddhist meditation lead to findings, and can we really call such findings scientific? Can we confirm the validity of the results of contemplative science, given that they are based on subjective experience?

THUAN: Whereas in science the primary methods of discovery are experimentation and theorizing based on analysis, in Buddhism contemplation is the primary method. Can you tell me if the word *knowledge* has the same meaning for a Buddhist as it does for a scientist? Is the kind of knowledge gained through meditation the same as rational knowledge? Must the contemplative lay aside the analytic approach to acquiring knowledge and purify his mind of the trappings of rationality? Must he silence thought so that he can grasp the Buddhist vision of reality?

MATTHIEU: According to Buddhist texts, the word *logic* (in Sanskrit, *pramana*) means "valid cognition." Logic is part and parcel of nearly all aspects of knowledge, including science and contemplation. But we do make a distinction between valid "conventional" knowledge and valid knowledge that is absolute, or ultimate. The former teaches us about the appearance of things—thus allowing us to tell the difference between a pool of water and a mirage, or a rope and a snake. But only the latter can allow us to comprehend the ultimate nature of phenomena (emptiness, the absence of inherent existence). They're both valid in their respective domains.

Logic and reason are used in analytical meditation, when we observe how thoughts work and examine the mechanisms of happiness and suffering. In this process we examine how our mind functions, such as the approaches it uses in order to perceive the world and make a mental picture of it. We also try to discover the mental processes that increase our inner peace and make us more open to others, as well as those processes that have a destructive effect. This analysis helps us to see how our thoughts are bound together, and how they then bind us.

As soon as meditation has inculcated in us an increased goodness and compassion, these enhance our reasoning abilities about life experience and help us to appreciate, for instance, the harmful consequences of hatred and the great advantages of being patient in our everyday lives. The training to cultivate emotions and ways of thought that are conducive to the pursuit of genuine happiness, and to free oneself from those that are detrimental to this pursuit, gradually transforms the stream of our thoughts, and eventually one's temperament. Love and patience aren't positive just by a priori definition, or because of some divine decree, but because they're the real causes of happiness.

T: Is the knowledge of enlightenment a higher level of knowledge than the rational knowledge of scientific thought?

M: There are a number of differences between enlightenment and conventional knowledge. First, enlightenment isn't only knowledge of apparent reality, but of the essential nature of reality. The false divide between subject and object vanishes, and reason is replaced with direct, clear, enlightened awareness, which mingles with the ultimate nature of phenomena until it is united with it. Far from being illogical, this type of knowledge has a perfect logic, based on the understanding of emptiness, which transcends the conventional logic of linear thought.

T: Can this knowledge be described as "intuitive" or "mystical"?

M: Terms like "intuitive" and "mystical" may cause confusion. If by intuition you mean a direct, immediate knowledge, then you're not far from the mark. But if intuition suggests a vague feeling about something unverifiable, or a hazy impression thrown up from our subconscious, then this definition reflects the delusion of our ordinary ways of thought. Rather than describe enlightened thought as mystical, we would say that it is the product of a non-dual, intimate union with the nature of the mind, which is clear, luminous, and concept-free.

Certainly, during meditation, one might have what could be called mystical experiences, but we must be careful about such transient events. If they don't improve our understanding of the ultimate nature of the mind, they lead us astray rather than enlighten us. Instead of aiming for ecstasy, or slipping into passive expectation, the point is to push analytical meditation to its extreme, so that our minds can rest in a state of luminous simplicity, over and above concepts. Then the realization of the mind's ultimate nature becomes vast, profound, and changeless. We reach the very source of thoughts, and observe what remains when they have gone. This state is, by its very nature, indescribable. By this I don't mean that it is too obscure to be described, although it's true that words are as powerless to describe it as they are to explain color to a blind man. For someone skilled in contemplation, nothing is clearer than the pure awareness of a mind free of conceptual thought.

T: Doesn't Buddhism use metaphor and allegory to describe the insights of enlightenment for this reason, because conventional language is so limited in its capacity to express these ideas? Doesn't it use disconcerting propositions, such as the koan of Zen Buddhism, as a way to help those who would learn those insights switch off the voices of logic and reason and leave the beaten track? A koan is a metaphysical riddle used to open up a student's thoughts. For example, when, after clapping his hands, a teacher asks, "What is the noise made by one hand?", this question is intended to break the train of discursive thoughts momentarily. In the interval between two thoughts, the disciple may then glimpse the true nature of the mind, which lies above mental constructions.

M: When trying to express the different degrees of understanding, the mind's ultimate nature and the emptiness of phenomena, we are often lost for words. But this does not involve abandoning reason; rather we transcend it. We say that it's as hard for a contemplative to express in words his understanding of the mind's nature as it is for a mute person to describe the flavor of honey. That's why we often fall back on images, which are never perfect, but which do reveal some aspects of spiritual realization, as though pointing at the moon with your finger. But you must then look at the moon, not the finger!

In this approach, a spiritual teacher can sometimes use quite unexpected means to break our conceptual habits and show us the freshness of a mind free of mental constructs. On a bright autumn night, on the mountain slopes that overlook Dzogchen Monastery in eastern Tibet—a marvelous place, where I was fortunate enough to stay—a nineteenth-century Tibetan hermit named Patrul Rinpoche was sleeping outside with one of his disciples. He suddenly called to him, "Didn't you tell me that you still don't know the true nature of the mind?"

"That's right."

"But it isn't hard."

He then asked him to lie down beside him. The disciple, whose name was Lungtok, lay down on his back and stared up at the stars. Patrul Rinpoche went on, "Can you hear the monastery dogs barking?"

"Yes."

"Can you see the stars shining?"

"Yes."

"Well, that's meditation."

At that very moment, Lungtok intuitively understood the nature of the mind. The cumulative effect of years of meditation, his teacher's presence, and this special moment triggered off that inner understanding.

By its very nature, ultimate knowledge—enlightenment—lies beyond any concepts. All other types of knowledge are incomplete. A theory can describe only a part of reality, because it uses propositions that are limited by the very nature of conceptual thought. Doesn't this limitation to theoretical science remind you of the famous Incompleteness Theorem of the Austrian mathematician Kurt Gödel?

T: Gödel's theorem does indeed imply that there are limits to rational thought, in mathematics at least. This theorem is generally considered to be the twentieth century's most important discovery in logic. In 1900, the German mathematician David Hilbert challenged his colleagues to devise a general procedure for determining whether any given arithmetic proposition is true or false. Doing so would put all of arithmetics (and, later, all mathematics) on a consistent logical basis. Kurt Gödel took up the challenge, but not in the way Hilbert had intended. In 1931 he published what is perhaps the most extraordinary and mysterious theorem in mathematics. It showed that any coherent arithmetic system must contain propositions that are "undecidable"—that is to say, mathematical statements that can't be proved or disproved logically. What's more, one cannot prove the coherence of that system without going outside of it and adding supplementary axioms. Thus any such system is intrinsically incomplete, and hence the name "Incompleteness Theorem."

Gödel's proof of this theorem certainly caused a stir in the world of mathematics.[1] He had shown that logic is fundamentally limited and that Hilbert's dream—to come up with a rigorous proof of the overall coherence of mathematics—was doomed to failure. The theorem has also had huge repercussions in other fields, such as philosophy and computer science. In philosophy because the power of rational thought has been shown to have limits, and in computer science because Gödel's theorem means that there exists mathematical problems that cannot be solved by a computer.

M: Buddhism has long argued that linear thought and discursive logic have intrinsic limitations. The path of enlightenment doesn't reject rational thought, but rather transcends those limitations. Reason is insufficient to express the ultimate truth, because there is a fundamental limitation in the structure of reasoning that prevents it from attaining direct knowledge of the Absolute.

T: Gödel's extraordinary theorem reveals a natural limit to scientific knowledge. In order to move past this limit, I agree that we need to call in other approaches, such as the one taught by Buddhism.

M: This is a vital point, because so many people invest science with an almost mystical power. They see it as the means to one day answer all questions. But this is far from the case. In fact, what we generally call "science" is incapable of describing most of what we experience.

T: Science is also not nearly as objective in its analysis as the ideal description of the scientific method suggests. For one thing, scientists are influenced in how they interpret their results by their professional training—their apprenticeship with teachers, interactions with colleagues, reading published work. Thus, once carried out, observations and experiments are analyzed and interpreted according to each scientist's inner world of concepts and theories. For example, astrophysicists will turn to a theory explaining how galaxies are formed, while their physicist colleagues will call upon one describing nuclear forces. Adopting one theory rather than another is also sometimes a question of bias. Researchers are influenced by the opinions of their teachers or colleagues (what we call a scientific "school") or, even worse, by fashionable ideas. For in science, as everywhere else, fashion must be treated with caution. The theory with the greatest number of supporters isn't necessarily the right one. Most supporters won't have examined it critically, and accept it out of conformity, or else intellectual laziness, or even because it has been defended by certain prominent, eloquent voices.

A scientist can't observe the world in a purely objective way. Einstein wrote:

> *Physical concepts are free creations of the human mind, and are not, however it may seem, uniquely determined by the external world. In our endeavor to understand reality, we are somewhat*

> *like a man trying to understand the mechanism of a closed watch. He sees the face and the moving hands, even hears its ticking, but he has no way of opening the case. If he is ingenious, he may form some picture of a mechanism which could be responsible for all the things he observes, but he may never be quite sure his picture is the only one which could explain his observations. He will never be able to compare his picture with the real mechanism, and he cannot even imagine the possibility of the meaning of such a comparison.*[2]

When several plausible but incompatible theories are put forward to explain one phenomenon, our choice among them often depends on our viewpoint. Einstein was never able to accept quantum physics' probabilistic description of the atomic and subatomic world because of his attachment to realism. He spent years vainly trying to find a flaw in the logic of quantum theory. As a consequence, he moved away from particle physics and expressed only a small interest in the great discoveries that revolutionized this subject in the 1950s.

M: Scientific theories are also deeply influenced by scientists' metaphysical viewpoints. Western researchers tend to believe that there's a solid reality behind the veil of appearances. Researchers who have thoroughly explored Eastern culture have less difficulty in calling into question the solidity of the real world. They are more open to the idea of interdependence. Scientists inherit a way of thinking from the culture they grow up in.

Alan Wallace, the philosopher of science who was also a translator of Buddhist texts, wrote as follows about this:

> *Upon confronting a diversity of theories that equally account for the same body of experimental evidence and that yield identical predictions, we may become disheartened and lose interest. Or we may assume that only one of those theories (or an unformulated one) represents physical reality. . . . The belief that there is an objective physical reality that can finally be represented by one and only one theory is a metaphysical*

> *assumption held by many scientists today. . . . If the presence of multiple incompatible theories accounting for the same phenomena is common in physics, what does science have to tell us about the nature of the objective universe? It would seem at this point that physics as such presents us with a number of options, among which we can try to choose the realistic one on the basis of our own metaphysical predilections! . . . Can we place a limit on the number of possible theories to account for a single body of evidence? Who can put a limit on the creativity of the human imagination, or on theories that lie beyond our imagination? . . . Moreover, if our choice of an account of the universe is ultimately a metaphysical one, why should we limit the possible choices to those presented by science?*[3]

T: But the prejudices of scientists are not all bad. These prejudices can, in fact, be important in inspiring scientists in their work. For if they had no preconceived ideas, or to use philosopher of science Thomas Kuhn's words, no paradigm,[4] how could scientists pick out the information most likely to be meaningful, and most suggestive of new laws and principles, among all the data that nature throws at them? This process of sifting is an essential part of the scientific process. The greatest scientists are those who have mastered the art of going straight to the essential while leaving aside what is insignificant. We've seen, for example, how Newton focused on linear, nonchaotic systems in order to fashion his theory of universal gravitation. Scientists see only what they can see, or what they wish to see.

M: Einstein also said, "On principle, it is quite wrong to try founding a theory on observable magnitudes alone. In reality the very opposite happens. It is the theory which decides what we can observe."[5]

T: Charles Darwin, the father of the theory of evolution, had a revealing story to tell about that. During his travels he spent a whole day on a riverbank and noticed nothing special, nothing but pebbles and water. Eleven years later he returned to the same spot, but now, owing to his subsequent studies, he was expecting to find evidence of an ancient glacier. Sure enough, this time, the evidence was blindingly obvious. Not even an extinct volcano could have left more visible traces of its past activity than this ancient glacier. Darwin only found what he was looking for when he knew what he was looking for. There are countless similar examples.

M: Scientists also tend to fit new facts into preexisting conceptual models and avoid calling into question the fundamental precepts of the field they're working in.

T: Yes, but, that said, sometimes when new facts turn up that don't fit into an existing framework, a scientific revolution, or paradigm shift, is kicked off.[6] This also happens when geniuses spot connections between phenomena that were previously thought to be separate. Norwood Russell Hanson, a historian of science, remarked, "The paradigm observer is not the man who sees and reports what all normal observers see and report, but the man who sees in the familiar objects what no one else has seen before."[7] Newton understood gravity when he saw the link between an apple falling to the ground and the motion of the Moon around the Earth. Relativity became clear to Einstein when he grasped the interconnection between time and space. But such imaginative achievements don't happen purely by chance. They result from years of learning and thought.

I don't believe that the lack of objectivity inherent in the scientific method means that science as a whole is fundamentally flawed. Science has a defense mechanism, which always ends up putting it back on the right track, even if it does stray down blind alleys from time to time. This mechanism is the constant interaction between theory and observation. There are two possibilities: either new observations and the results of recent experiments agree with the current theory, and therefore reinforce it; or they do not, in which case the theory must be modified or else replaced by another one that describes better the observations or experimental results. The new theory usually involves a number of new predictions, which scientists then seek to verify. The new theory will be accepted only if its predictions are confirmed.

In addition, the observations and measurements that corroborate it must be reproducible and confirmed independently by other researchers using different techniques. This is a fundamental point, especially when it comes to discoveries that call into question widely accepted theories and, in Kuhn's words, "alter the paradigm." Researchers are naturally conservative. They don't like new theories springing up and, from one day to the next, sweeping away all their hard-won knowledge. This is just as well, because in science it isn't enough to destroy, we must build anew. And it's difficult to rebuild on ruins.

The experimental method involves a constant to-ing and fro-ing between observation and theory. It allows us to slowly approach an accurate description of phenomena, even if we do sometimes take wrong turns that lead to dead ends and have to go back to square one. Science doesn't progress in a straight line, as it is so often depicted as doing, but in a zigzag.

M: This to-ing and fro-ing between theory and observation allows us to check that a theory explains and predicts certain facts correctly, but it doesn't call into question the researchers' metaphysical prejudices. One researcher can easily prove to a second one that he was wrong about the life span or mass of a particle, but that doesn't stop both of them from still being convinced that the particle really exists. During each scientific revolution, researchers think that they've finally got a true picture of reality. The unfounded certitude that goes with this illusion is also a philosophical prejudice. In 1939, Alfred North Whitehead commented, "Fifty-seven years ago it was when I was a young man in the University of Cambridge. I was taught science and mathematics by brilliant men and I did well in them; since the turn of the century I have lived to see every one of the basic assumptions of both set aside. . . . And yet, in the face of that, the discoverers of the new hypotheses in science are declaring, *'Now at last, we have certitude.'*"[8]

T: Some scientists' intellectual arrogance has even led them to make grandiose pronouncements about the end of science—we have now understood everything, there's nothing left to discover! So far, history has always proved them wrong. At the end of the nineteenth century, Lord Kelvin, the great specialist in thermodynamics, declared that the study of physics was over, and that all that was left for future generations was to refine measurements and add more decimal places. He could hardly have been more off the mark. A few years later, physics was overturned by relativity and quantum mechanics.

In my opinion, we will never know the whole truth if we limit ourselves to the scientific approach. We'll continue getting nearer the goal, but we'll never quite reach it. The melody will remain forever secret.[9]

M: But isn't this pursuit of complete scientific knowledge ultimately misguided? Are we going to persist in the vain attempt to grasp some hypothetical reality that, as quantum physics shows, slips away from us as soon as we get too close? Or are we going to aim for an ultimate knowledge, like enlightenment in Buddhism?

T: Personally, I don't really agree with Kuhn, who rejects the idea that there's a "goal" that science nears in measured steps. I do think there is a goal. It's the complete and detailed knowledge of inanimate and animate phenomena. Science is nearing this goal and making definite progress. Our knowledge of the world is infinitely richer than it was during the Renaissance. But this progression doesn't follow a straight line; it veers off into many detours and reverses. So I do believe that science is making remarkable progress, but, that said, I concede that science does not have the means to reach the sort of ultimate knowledge that Buddhism talks about.

You often describe Buddhism as a contemplative science. What do you mean by "science" here? Can the method I've just described be used in the contemplative approach?

M: What I mean by "science" is knowledge that is rigorous, coherent, verifiable, and doesn't study only physical phenomena, but the whole scope of our living experience. Why limit the word "science" to what can be checked using instruments or equations? The discovery of the mind from within is not akin to discovering a phenomenon to be weighed and measured. A science must be tested, confirmed in experiments, and available to all experimenters. The last point doesn't mean that just anyone should be able to do science, as easily as people switch on their TV sets. This isn't the case for the natural sciences, or for contemplative science, either. Physicists need years of study before they can understand the equations of the universe. It also takes contemplatives years to acquire correct knowledge and a lasting control over their minds. So it would be wrong to argue, "You contemplatives, you claim to have a close understanding of consciousness, but how can I know, given that I can't verify your claims?" Most people who speak like that would also be incapable of explaining at a moment's notice why they accept the results of special relativity or the calculation of the structure of the atom. In both science and contemplation, direct verification involves training using reason and experience. The important thing isn't to get immediate results, but to be able to get there using adequate, verifiable means. A science's validity rests on the results obtained by researchers who have consecrated enough time and energy to confirm their hypotheses. If they agree, it's reasonable to believe them and call their knowledge "science." Contemplative science is based mainly on personal experiments, and only those who devote themselves to conducting such experiments can really share it.

T: But isn't there an important difference between the intersubjective knowledge of Buddhist contemplation and the objective knowledge that stands at the base of natural sciences? The first position comes down to saying, "Stand where I'm standing, and you'll see the same thing I see," while the latter says, "No matter where you stand, if we look toward that point, then we see the same thing." A good example is Müller-Lyer's optical illusion, using two arrows:

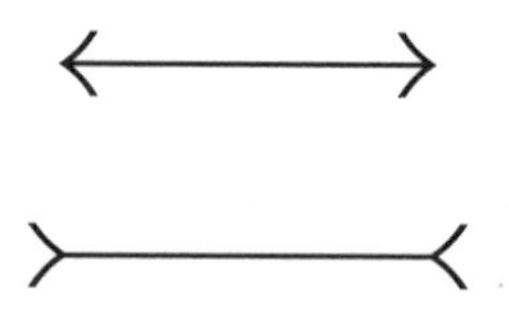

Here, objective knowledge is obtained by using a ruler: anyone who reads its gradations will see that the two lines are the same length. Intersubjective knowledge comes from questioning people about how they see the two lines. We observe that they all agree that the first arrow is about 5 percent shorter than the second one. In both cases the agreement is overwhelming, but in the first instance, a common object was used (a ruler), and in the second, people used their experience.

M: This doesn't mean that Buddhist contemplation isn't scientific. If we define the terrain field of science as what can be physically studied, measured, and calculated, then right from the start we leave out everything that is experienced in the first person, and all immaterial phenomena. If we forget this limitation, then we soon start affirming that the universe is *everything* that can be objectified in the third person, and *only* what is material. Consciously or unconsciously, we've adopted a metaphysical position.

T: This is unfortunate. We then run the risk of missing out on important discoveries. On the other hand, we had to exclude the immaterial in order to make the natural sciences progress. If contemplative Buddhism is truly a science, then what hypotheses do Buddhist researchers have?

M: In our science, because our goal is to discover the means to put an end to suffering, we start by examining the mind in order to see what leads it to a state of profound satisfaction and what destroys its serenity. We have perceived that feelings such as malice, jealousy, lust, and envy never produce long-lasting happiness. They come from egocentric compulsions, and make us desire anything that seems pleasant to us, and reject anything that seems unpleasant. They push us into an illusory quest for happiness, which causes only suffering. If we realize our mistake, then we see that we must transform the negative impulses that cloud our judgment. Our working hypothesis is thus as follows: Suffering comes from afflictive thoughts, or mental toxins, which themselves come from our attachment to our egos. Such attachment is the basis for self-centeredness, in which one regards our well-being as being intrinsically more important than the well-being of others. By unmasking this deluded attachment, we can free ourselves little by little from the cause of suffering.

T: What happens at the experimental stage?

M: The experimental stage is when we analyze the ego's characteristics until we understand that it's just a mental label that we attach to a dynamic stream of consciousness in perpetual transformation. We thus understand what happens when we dissolve completely our attachment to the notion of "ego." Then we also observe the beneficial effects of positive thoughts, such as generosity, patience, and love, and the ill effects of their opposites. Slowly we begin to understand the laws that govern them. We can then examine the various methods that can be used to free the mind from these mental poisons, and so start practicing them.

T: Are these propositions really "laws," as we've just defined the word—that is to say, statements describing necessary and constant relationships between phenomena?

M: They work coherently. For example, hatred *never* produces genuine happiness in the long term. Some people may experience a sadistic pleasure in moments of hatred, but we all know that it's impossible to live in peace with this sort of feeling. Its mechanisms are determined by causality. Anger and jealousy inevitably eliminate joy, while love and compassion engender it. No matter how hard we try to ignore this truth, we can never escape from its consequences. This is not an abstract approach, but an experimental investigation bound up with a profound reflection—as long, methodical, and rigorous as that of a scientist analyzing the laws of physics and mathematics. This reflection doesn't lead to equations, but it finally makes the mind limpid, stable, and calm.

T: So you start by observing others, and then later turn to observing yourself?

M: Both go together. While observing others might open our eyes, the vital thing is to turn our gaze inward and observe our own minds. Even if we can go on deceiving others, it becomes harder to hide the truth from ourselves. That's why we must constantly examine the mirror of our minds. Contemplative experiments destroy our preconceived ideas about the world and ourselves, and put us face-to-face with the true nature of things. They show us clearly that the ego is just a mental construct. This discovery has great repercussions on how we see the world and how we live.

T: How does Buddhism verify its findings? In science, observation and experimentation are objective; their results don't depend on who carries them out. Vietnamese and American physicists will get the same measurements as their French colleagues if the work is done correctly. That objectivity is the basis of the experimental method. A scientific result, particularly an exceptional one, is accepted only after it's been independently checked by other teams, using different methods and equipment. But Buddhist knowledge comes from meditation and introspection, which are, a priori, private and subjective. So how can such knowledge be universal?

M: For a long time there have been prejudices in the West against contemplative science, because people didn't know how to deal with it.[10] People claimed that the mind wasn't a reliable instrument and that experiments done on it weren't reproducible. But those critics hadn't devoted themselves to the process of personal experimentation.

Inner experiments have an undeniable use for those who carry them out, and the long-term effect can be judged objectively. Our way of being, of speaking, and of acting is transformed. We move toward altruism, serenity, tolerance, and inner strength, which are the main criteria of success in this sort of experiment. It's certainly true that an outside observer can't directly check the effect of a given Buddhist practice on my mind. But he can check the result on his own mind, if he bothers to make the effort.

T: What procedure do you follow?

M: In physics and astrophysics, scientists can use ever more powerful instruments. In the contemplative method, the only instrument is the mind. To begin with, it's badly adjusted, capricious, inconstant, and disoriented. It's as difficult to calm as a wild beast caught in a net. So you must adjust it and widen its field of vision, just as you enlarge the diameter of a telescope. This training isn't an end in itself, but a crucial step in honing the instrument of introspection. A sustained effort makes the mind more stable, calm, and manageable. You eliminate the waves of gross emotions, then the tossings of mental agitation and discursive thought. You identify the basic mechanisms of attraction and repulsion, of the clouding or illuminating of the mind, and of inner slavery or freedom.

T: I'd like to know if you must go past this purely analytical meditation, and how this is achieved. Also, if we identify the thoughts that disturb us, is this the same as neutralizing them?

M: You mustn't try to block them, but instead go back to their source and examine their basic nature. You then see that thoughts aren't intrinsically alienating. Rather, it is the attachment we feel to those thoughts, the difficulty we have letting them go, that obscures the fundamental nature of mind. By examining thoughts, you see that they have no shape, no color, no position, and that they fade away under scrutiny. They come from nowhere, and have nowhere to go when they vanish. Their apparent solidity melts like frost in the sun. We can then remain in the mind's primordial simplicity, the natural clarity of the present moment, the immutable serenity of the mind's ultimate transparency, without summoning up the past or imagining the future, and without hope or fear.

This exercise would be of little use if it were not constantly repeated with a view to grasping the elusive nature of our thoughts. An understanding of their emptiness releases us from their hold. Disruptive thoughts gradually lose their power to whip up storms inside us and to make us negative as regards others. Little by little, we become expert at this liberating process. When thoughts appear, we watch them come and go, like an old man quietly watching children at play.

T: How long does it take a normal mortal to reach this stage? An entire lifetime?

M: Not necessarily. The time depends on our faculties and perseverance. At the beginning, recognizing your thoughts as they arise is like spotting a familiar face in a crowd. Later on, thoughts free themselves, like a snake undoing a knot in its body without any assistance. Of course, this "freedom" has nothing to do with giving free rein to your every whim. In this context, freedom means that your thoughts stop running together and dragging your mind down into delusion. Finally, the third step is perfect mastery of this freedom from thoughts, which can no longer do you any harm. They're like a thief in an empty house. The thief has nothing to gain, and the owner has nothing to lose. Thoughts arise, then dissolve without leaving a trace, like a drawing sketched on water.

T: But, unlike scientific experiments, which must be reproducible, doesn't this experience vary greatly from one person to another?

M: Of course, personal contemplative experiments can't be directly witnessed by a third party, as is the case for ordinary scientific experiments. And at first they don't produce objective results. Contemplatives sometimes run the risk of incorrectly gauging the worth of what they're doing. As I've already said, however, success leads to a lasting change in the person, which can be judged objectively.

What's more, the goals reached through such introspection are remarkably similar—inner peace and strength, nonattachment, loving-kindness, and so on—despite differences in personality. The means and techniques used are extremely similar, and the texts give detailed descriptions of the different stages of the journey. Some are more gifted than others at following this sort of discipline, and arrive at a higher level of mastery over their minds—you might say that some people build "mental telescopes" with a diameter of one meter, and others of ten meters—but everyone who travels this path gets consistent results.

T: Do descriptions by different authors agree?

M: We mustn't forget the difference between objectivity and intersubjectivity. When done properly, the contemplative approach has led to an astonishing intersubjective agreement over the centuries and generations of followers. The descriptions don't always use the same images, but the stages in the progression and the results are the same. For example, some authors say that thought is initially like a frothing waterfall, then like a stream with occasional eddies, then like a large river with the odd ripple running over it, and finally like the ocean, whose depths are never disturbed. There are many books with more-technical descriptions and with numerous details that can be checked—if and only if, as I've said before, you're willing to take the trouble. These texts also describe the stages of nonconceptual meditation and of the pure contemplation of the mind's nature, at the end of which you arrive at enlightenment, the ultimate state of inner knowledge.

T: That of the Buddha?

M: Of the Buddha and those who have followed him. Of course, there are intermediate stages of spiritual development, which provide a considerable sense of fulfillment when reached. It's said that the Buddha's enlightenment is greater than that of a traveler setting out, in the same proportion as the heavens are bigger than what can be seen of them through the eye of a needle. But in both cases, what you see is the sky. So, without having reached ultimate enlightenment, you can discover some of its qualities. In general terms, we might say that contemplative science is basically qualitative, while physical science is basically quantitative.

T: Certainly, but to come back to enlightenment, isn't the Buddha the only person to have reached it? Can we all succeed?

M: The Buddha affirmed that everyone who correctly follows the same path as he did will have the same result. Did he not say, "I have shown you the path, it is up to you to follow it"? Each being can potentially reach a perfect understanding of nature and the mind. This is, in a way, Buddhism's "original goodness." If you free yourself of all the negative thoughts that cloud your mind, you will experience indestructible peace and compassion.

T: But how do you know if you've gone up a blind alley? In the natural sciences, we confront a theory's predictions—a planet's orbit, for instance—with observations. If they agree, then the theory can be accepted. If they don't, then it must be amended or rejected.

M: If scientific information is a geographical map, then the Buddha's teachings are like a travel guide. The farther you go, the clearer they get. You notice that if you diverge from the guide's instructions, then you run up against obstacles that slow down your progress. This can lead to discouragement, doubt, confusion, or aversion. But in the right hands, obstacles can be used as catalysts so that progress becomes even faster. All these different possibilities have been analyzed with great precision in various textbooks.

T: This is starting to sound like the natural sciences' to-ing and fro-ing between theory and experiment. Here, the theory is that attachment to the ego is the source of all our troubles. The method is an analysis of the ego and its effects. The experiment is an application of the method by introspection and contemplation. Then the result is the elimination of the attachment and the afflictive emotions that it creates. If we run into obstacles, then other contemplative tools are used to get around them. We move to and fro between various methods in order to get rid of the ego, until we discover which one is best suited to a particular person. I now understand better why you used the term "contemplative science." Buddhism's techniques for reaching enlightenment are fundamentally similar to science's methodology. What surprises me more than anything else is that introspection can be reproducible.

M: Psychologists who have studied introspection have generally failed because of a lack of sustained training and a refusal to take into account the experiences of ancient traditions. They've rashly concluded that experiments such as those that Buddhists do aren't reproducible. Introspection has also been seen as suspect by the natural sciences because, until recently, its approach has been purely qualitative. When you start a new subject in natural science, the first thing you want is data, graphs, images, and so on.

T: But neurologists are now making progress in measuring some of the kinds of mental processes you study in Buddhism, aren't they?

M: Indeed, mental imaging has progressed enormously. We can, for example, distinguish between the areas of the brain that are active when we make a given gesture, and those that are active when we think about the gesture. In the same way, the area changes when we hear an abstract word or a concrete word. Recently, Francisco Varela and his team have shown how the different parts of the brain are linked together during the recognition of an object.[11] Now other researchers, too, such as Richard Davidson and Paul Ekman, would like to develop a study program about the neurophysiology of meditation.[12] But will neurology ever be able to describe pure introspection and the observation of the mind's ultimate nature, which is essential for contemplatives? At best we will detect different cerebral activity, but that won't tell us much about the *quality* of the meditation, just as working out which distinct cerebral activities match the identification of red and blue doesn't tell us anything about how those colors are experienced.

T: It must be said that we're still far from understanding what happens in our brains when we love, hate, create, or feel joy or sorrow. We should be careful that the cognitive sciences' imitation of the natural sciences' quantitative approach doesn't lead to excesses. This is just what happened at the beginning of the century with behaviorism. In its attempt to promote psychology as an "exact" science, it tried to study the behavior of living creatures only by observing their responses to external stimuli. By rejecting everything that can't be observed directly, behaviorists denied the very existence of the mind, which is absurd.

M: Even if no measurable indication were found in subjects engaging in various kinds of meditations (although preliminary experiments indicates that there are significant differences), that wouldn't disprove the validity of this inner experiment and its power to change us. On the other hand, scientists can have any sort of good or bad qualities. That changes nothing about the results of their chemistry experiment or the measurement of the wavelength of stellar light.

The essential aim of contemplative science is to become a better person. The way of life that this implies can seem off-putting. Finding the inner energy to get rid of all our faults is no easy job. The idea of attacking our own egos is repulsive. We then slump back into an inertia that is one of the main obstacles to the spiritual life.

T: Do Western cognitive sciences and psychology study the same subjects?

M: Psychology examines feelings, emotions, behavior, memories, in fact all the mechanisms that condition our conscious states. The cognitive sciences try to explain the mental processes associated with perception, remembering, learning, and so on. But, despite becoming increasingly interesting, their principal aim still isn't to transform people. So it all depends on the motivations of the scientists who conduct such researches.

T: Can't psychology and psychoanalysis help us to reach that goal?

M: In theory, psychology could do so, but it would have to widen its scope and start to use certain meditation techniques. As for psychoanalysis, the aim is different. Its purpose is to set up a compromise, a stabilization, a status quo, which is acceptable to the ego, and so return to a "normal" state of affairs. It's a question of finding a balance between the impulses being played out in the ego and what is socially acceptable.

Contemplative science, on the other hand, aims to dissipate our illusory egos. In psychoanalysis, the ego becomes our main preoccupation. We even strengthen it, and so find ourselves engulfed in the illusion of this ego, which we manipulate in every possible way, like a piece of sticky paper that we shift from one finger to another without being able to get rid of it. In contemplative science, we burn that illusion, like feathers that leave no trace. Thus the aim lies far above a stabilization and balancing of our ordinary state.

Enlightenment isn't the normalization of our disruptive emotions, and it is certainly not the reconstruction of the ego. It implies a total freedom from their grip. It also features an inner joy and an unbreakable plenitude, which seem to be lacking in psychoanalysis. There's a complementary side to contemplative science. Not only does it allow us to comprehend the mind's nature, but it also lets us hone our understanding of the world of phenomena, thanks to the interdependence between our consciousness and the world it perceives. Also, the role of the psychoanalyst is very different from that of the spiritual teacher. The latter is a model, a living example of what we could become.

T: For scientists, intellectual delight comes from discovery. It's extremely exciting when a small part of the veil hiding nature's secrets lifts to reveal what was previously unknown. But that isn't enough for a fulfilling life. These moments when the truth is revealed are magic, but extremely brief. Since the birth of modern science in the sixteenth century, we've experienced an exponential growth in our knowledge, but this hasn't made us better people. Contemplative science could help us to attain true wisdom. This is becoming all the more urgent now that we have the means to disturb the ecological balance of the entire planet and even destroy ourselves with our nuclear weapons. The ethical problems we're faced with are becoming increasingly pressing, particularly in the field of genetics, and meanwhile the gap between the rich and the poor keeps widening. . . .

14 Beauty Is in the Eye of the Beholder

Is there a notion of beauty in scientific investigation and the theories that guide it? What is beauty in Buddhist terms?

THUAN: Often the theories that describe nature best and are closest to experimental observations are also what scientists would call the most beautiful. The notion that scientists would use such a word as "beauty" may strike readers as odd, because scientific work is generally perceived as being cold, rational, and lacking any aesthetic motives. But scientists have always spoken of beauty. Let's listen, for example, to the eloquent words of the French mathematician Henri Poincaré: "Scientists do not study Nature for utilitarian reasons. They do it because they find it pleasurable; and they find it pleasurable because Nature is beautiful. If Nature were not beautiful, it would not be worth studying, and life would not be worth living."

We have no trouble perceiving the intrinsic beauty of natural phenomena, such as that of roses, sunsets, stars, and galaxies. I am filled with wonder every time I see images relayed by the telescope to my computer screen, images of young stars in a nebula, or the exquisite shape of the spiral arms of a galaxy that's several million light-years away. But in addition to this visible beauty, there is the subtler and more abstract beauty of theories.

One reason that a theory is described as beautiful is that it is recognized as being inevitable, necessary, and is immediately seen as self-evident once it's been fully worked out. When presented with a new theory, physicists say to themselves, "It's so beautiful it must be true. Why didn't I think of it?" In this way, Einstein's Theory of Relativity is as beautiful as a Bach fugue, not a single note of which can be altered without ruining its harmony, or the *Mona Lisa*'s smile, not a trait of which can be changed without destroying its balance.

Another chief hallmark of a beautiful theory is its simplicity. I don't necessarily mean that the equations are easy, but that the underlying ideas are. Copernicus's heliocentric universe, with the planets revolving around the sun, is simpler than Ptolemy's geocentric version, in which the Earth held the central place, while the planets moved along circles (called "epicycles") whose centers themselves moved around yet other circles. The Copernican model is beautiful because it describes the motions of the planets much more simply. A beautiful theory has no unnecessary frills. It passes the test of Occam's razor: "What is not strictly necessary should not be used."

Finally, the last and most essential quality of a beautiful theory is its truthfulness. The final judgment of its validity rests on how well it conforms to nature and whether it reveals previously unsuspected relationships.

MATTHIEU: What sort of "truthfulness" are we talking about here? When you say that the theory conforms to nature, you mean that it does so empirically, right? But as we've discussed, scientific experiments can't reveal the ultimate nature of reality.

T: What I mean is that the theory conforms to the truth as shown by our apparatus, or what Buddhists would call "conventional truth." For instance, let's take Einstein's General Theory of Relativity. According to most physicists, it's the most beautiful and harmonious intellectual edifice that the scientific mind has ever built. Not only did it bring together fundamental concepts of physics that had been viewed as totally unrelated—such as space and time, matter, energy and motion, acceleration and gravity—but it also revealed extraordinary phenomena that were previously unknown. Even now, General Relativity still surprises us with its hidden riches. In 1915, when the theory was published, we thought that the universe was static. In fact, Einstein's equations showed that it must be dynamic—either contracting or expanding. The physicist didn't trust his theory enough; otherwise he could have predicted that the universe was expanding fourteen years before Hubble's discovery.

Black holes are another example of a phenomenon predicted by relativity. Once again, Einstein didn't believe in their existence. He said that nature abhorred singularities such as black holes, and that relativity was incapable of describing them. There, too, he should have had more faith in his theory, because since then black holes have been detected in the Milky Way and in other galaxies, too.

A third example is that of what's called gravitational lensing. General Relativity tells us that in certain places the gravity of massive galaxies bends space and deviates the light coming from distant objects, thus creating "cosmic mirages." Such galaxies are called "gravitational lenses" because, like a lens, they refract and focus the light passing near them. They were discovered in 1979.

Inevitable, simple, and true to reality—those are the characteristics of a beautiful theory!

M: I'd say that conforming to truth comes pretty close to the Buddhist idea of beauty. But what we mean by "truth" isn't the conformity with external phenomena, but rather conformity to human beings' profound nature.

The simplest definition is that beauty is what gives us a feeling of plenitude. According to the circumstances, this can be expressed as mere pleasure or as a deeper sense of happiness. There are different levels of beauty matching different levels of plenitude. Something that gives us a passing moment of happiness can be called relative beauty, while absolute beauty contributes to long-lasting or even unalterable fulfillment. Spiritual beauty, for example that of a Buddha's face, is particularly rich because it allows us to sense that enlightenment exists and that we can reach it.

But beauty can be perceived in quite different ways by different people and societies.

T: Yes, the perception of beauty depends on a number of cultural, social, psychological, and even biological factors. The ideal woman in Renoir's time was decidedly buxom. In the 1960s, the waiflike Twiggy was considered an icon of beauty. The painter Vincent van Gogh died in poverty because he couldn't sell his paintings, but half a century later people were paying a fortune for them. Appreciation of a scientific theory, however, is far less dependent on cultural context. Physicists all over the world appreciate the beauty of General Relativity in the same way.

M: But that's because they've had a similar education. I don't think that a member of a primitive tribe would ever call relativity beautiful!

Beauty can also be seen as harmony between the parts and the whole. In Buddhist art, there's an extremely precise iconography that describes the ideal proportions in the drawing of a Buddha. A grid is used in order to place the curve of the eyes, the oval shape of the face, and various parts of the body with great precision. These features correspond to perfect harmony and are external reflections of enlightenment's inner harmony.

T: I've always been struck by how different representations of the Buddha, be they drawings or sculptures, have a beauty and balance that invariably give off a deep feeling of serenity and soothe the mind.

M: So beauty varies according to how each of us conceives aesthetic pleasure, and can range from the accessory to the essential. All thinking beings share certain profound conceptions of happiness and plenitude. Love and altruism are beautiful, while hatred and jealousy are ugly. Just look at the way the former beautifies a face, while the latter disfigures it. True beauty thus conforms to mankind's deep nature.

In Buddhism, this nature is defined as intrinsic perfection, full of love and understanding, and absolutely beautiful. The closer we come to our ultimate nature, the more we discover the inner beauty we all have. Ultimate beauty is perfect agreement with the Buddhahood, supreme knowledge and enlightenment. When we see a noble being, a radiant spiritual teacher, we intuitively know that we are in the presence of great spiritual beauty. Rediscovered harmony radiates from his or her face.

On the other hand, the characteristics of more-superficial relative beauty don't belong to the object itself, but are closely determined by the observer. Some people find a given object beautiful, while others find it ugly. An object is seen as being beautiful when it corresponds to our desires. Mathematicians are amazed by the beauty of an elegant equation, and engineers by the beauty of a machine. People in search of calm delight in listening to a Bach prelude. But a hermit contemplating the ultimate transparency of the mind has no need of such things. His harmony with the nature of the mind and phenomena lies elsewhere. For him, all forms are seen as manifestations of a primordial purity, all sounds as the echoes of emptiness, and all thoughts as the intertwining of wisdom. He no longer makes any distinction between the beautiful and the ugly, or the harmonious and the discordant. Beauty has become omnipresent, and plenitude immutable. It is said that "on a golden isle it is vain to search for ordinary pebbles."

15 FROM MEDITATION TO ACTION

Transform yourself in order to transform the world—such might be the motto of a practicing Buddhist. But how must we act on the world, and on what level? For contemplatives, a key question is how long they must continue the process of inner transformation before they should start acting on the world. Wouldn't they spend their time better by consecrating themselves to relieving others' suffering at once? Does Buddhism make enough effort in the field of humanitarian work?

THUAN: Buddhism believes that we should act on the world in order to relieve suffering, and surely such action should be as important to us as our spiritual development, shouldn't it? Wouldn't it be overly selfish to find peace and happiness just for ourselves, while everyone around us is suffering? The news is filled with reports of war, poverty, epidemics, and death. What's the use of a tiny island of plenitude amid an ocean of misery?

Some Westerners have perceived Buddhism as a passive, defeatist philosophy, which preaches that we should retreat from the world and learn to accept what happens to us and others, because we can't struggle against our karma. But isn't this idea wrong? Isn't compassion in fact at the center of Buddhist preoccupations?

MATTHIEU: At first sight, meditation and action seem to be diametrically opposed. On the one side we have contemplatives, who seem to act on the world only through meditation and prayer, while on the other we have people who are incredibly busy, whose actions are sometimes successful and sometimes less so, but which follow one another like waves. This frenzy often turns out to be rather inefficacious, given that it isn't based on any real personal transformation, or what we might call "spirituality" in the broadest sense. A lack of orientation and inner harmony means that our acts are often off-target. The benefits to society don't correspond to the effort expended.

We must build a bridge between contemplation and action. Experience shows that selfishness prohibits positive inner transformation. Such transformation can only come about through altruism. Plenitude cannot be reached by concentrating either only on ourselves, or only on the external world.

Compassion without action is indeed hypocritical; it brings cold comfort to those who suffer. So we must act each time we can and, even more, try to prevent suffering before it starts. Our own happiness is intimately linked to that of others. Most of our problems derive from the fact that we are not genuinely concerned with other people's well-being. Any happiness we feel that ignores others' unhappiness, or, even worse, bases itself on their suffering, can only ever be a pale imitation of happiness. As Shantideva said:

> *All the joy the world contains*
> *Has come through wishing happiness for others.*
> *All the misery the world contains*
> *Has come through wanting pleasure for oneself.*
>
> *Is there need for lengthy explanation?*
> *Childish beings look out for themselves.*
> *Buddhas labor for the good of others:*
> *See the difference that divides them!*[1]

Various texts dealing with the Buddhist contemplative life state that anyone who retires to a hermitage in the mountains just to escape from the problems of daily life is no better than the birds and the beasts that spend all their lives in remote places. Such renouncement brings us not an inch nearer enlightenment.

Apart from natural catastrophes, most human suffering is caused by malice, greed, jealousy, indifference—in fact, all the various aspects of egocentricity that stop us from thinking of others' happiness. One of Buddhism's fundamental practices is to think of others as being as important as yourself, to put yourself in their place, and finally to give them more importance than yourself.[2] A profound cure for our own egocentricity is a good way to limit the suffering of others.

But we must also distinguish between short-term remedies and long-term actions. Khyentse Rinpoche said:

> *When we think of all these beings suffering helplessly, we cannot help but feel tremendous compassion for them. Compassion by itself, however, is not enough; they need actual help. But as long as our minds are still limited by attachment, just giving them food, clothing, money, or simply affection, however vital this may be, will only bring them a limited and temporary happiness. If we wish to liberate them completely from suffering, we must first transform ourselves.*

A desire to act at once, with no preparation, is like wanting to operate on the sick in the street without first building a hospital. It's true that all the work required to build a hospital doesn't cure anyone directly, but once it's finished, people can be looked after far more efficiently.

The true contemplative realizes that he's incapable of reducing the suffering around him without first mastering a perfect understanding of the mechanisms of happiness and suffering. It's only when we've acquired sufficient inner strength that we can be of real use to others, by directly relieving their suffering, or by provoking changes in the society in which we live. Compassion is essential for any progression along the way of inner transformation. In the *Sutra That Authentically Resumes the Dharma*, we can read, "May he that desires to reach Buddahood not attempt many methods but one. Which? Great compassion. He that feels great compassion will know all of the Buddha's teachings, as though he held them in his hand." What's more, someone who has reached enlightenment spontaneously feels boundless compassion toward others.

So, from the beginning to the end of the Buddhist path, it's compassion that allows us to master hatred, greed, jealousy, and other mental poisons, and to put an end to the infernal cycle of suffering, both for ourself and for others.

T: Evil should certainly be rooted out. But isn't this vision too idealistic? Can any personal transformation be hoped for when it comes to such monsters as Hitler, Stalin, or Pol Pot? Shouldn't we use more-direct means to combat such evil?

M: Of course, it's unreasonable to suggest that such criminals, who seem immune to human feelings, can undergo personal transformation in the short term. Once psychopathic madness has reached a certain stage, it eludes reason, and then more drastic means are called for. But this in no way contradicts the validity of long-term remedies for diminishing the suffering in our world. By neglecting the importance of this long-term work, we become narrow-minded, like a doctor who prescribes painkillers but looks for no basic cure. In the course of history, we have seen certain cases, which are only too rare, of societies that have been changed by a desire for personal transformation. I'm thinking particularly about Tibet.

T: The Dalai Lama stands for nonviolence, and I admire the fact that he maintains this position despite the tragedy that has afflicted his country. He certainly possesses a great strength of purpose in not answering violence with violence. And I know that his attitude is attracting more and more sympathy throughout the world. But, unfortunately, the rub is that we are surrounded by countries and societies that don't adhere to pacifism, and so war and oppression can't be avoided. Must we, then, suffer without hitting back, thus running the risk of losing our country, culture, and lives? Some Tibetans have publicly questioned the Dalai Lama's policy of nonviolence, given that Chinese repression continues unabated, genocide has decimated the population, and there is a continuing attempt to eradicate Buddhist culture.

M: Using violence to free yourself as rapidly as possible from an unjust oppression, and make a lesser evil destroy a larger one, is risky. Violence generally leads to more violence. It's vital that we gauge all of the suffering that will result from a given situation. Force can be used only when it limits suffering, not when it produces it. The essence of nonviolence, even when one uses force, is to be totally free from malevolence and from seeking vengeance.

Pacifists who turn to violence may end up contradicting themselves and making declarations similar to the recent claim of a general in the Colombian army: "We want peace, but the only way to have peace is by destroying those who don't keep peace."[3]

On the other hand, if we consistently stick to a principle of nonviolence, then peace will probably be long-lasting when it finally arrives. Nonviolence isn't a passive approach, and it can be very effective. We should never forget the example of Gandhi, who mobilized his country through nonviolence. Not attacking aggressors with violence doesn't mean that we can't use all the other means—nonviolent but determined resistance, dialogue, political and economic firmness—in order to combat evil and reduce overall suffering. In fact, the worst caricature of nonviolence I can think of is the naiveté and complacency of Western heads of state as regards despots in general—and Chinese leaders, in the case of Tibet. Here we find a laxity and credulity that are absolutely indefensible, in terms of the nonrespect of human rights and international law. Similar weakness led European governments to close their eyes to the rise of Nazism in the 1930s, and for many Western intellectuals to flirt with communism after the war.

I'm sure that history will reserve a similar judgment for those who still tolerate the existence of *laogai*—the Chinese gulags. These forced-labor camps produce many of the manufactured goods that we unwittingly buy, and the toys we give our children for Christmas as what are meant to be gestures of love. But I suppose that the leaders of free countries are far less worried about the judgments of posterity than about making a courageous decision here and now.

T: I don't think this is because our leaders are naive. It's rather a perverse result of the globalization that everyone's talking about. Since all economies are now inextricably linked,Western leaders don't dare to stand up to China about Tibet. They're frightened they might lose a huge market of one and a half billion inhabitants. We mustn't forget that a quarter of the world's population is Chinese. But this realpolitik still doesn't justify the actions of our politicians.

M: So if we agree that we want to take positive action in the world, the first step is to acquire the *ability* to act. Powerlessness is the first thing people encounter when starting on this quest. They can't help themselves because they haven't worked out the mechanisms that govern happiness and suffering. So they are all the more incapable of helping others.

By relegating our inner torments to a lower level through steady practice, we acquire an inner peace that naturally makes us more sensitive to the suffering of others. By gradually understanding the global interdependence of beings, we start seeing the world differently and acting more justly. People who have put themselves at the service of others radiate harmony. You just have to see how being in the presence of the Dalai Lama for a moment brings out the best in people. I've seen countless journalists, blasé politicians, pretentious celebrities, and ordinary people who had no particular interest in Buddhism who have been transformed by spending half an hour in his company. Meeting someone who overflows with love and concern for everyone else's well-being completely bowls them over.

T: I have had the great good fortune to meet the Dalai Lama, and I can attest that he radiates such strength, serenity, and "tranquil will" that you can't help being profoundly moved. So that is your definition of Buddhist action—transmitting the harmony you have found yourself.

M: Yes, and in fact there is no other way. You can discipline other people's words and actions externally, but it's only from the inside that they can adhere willingly to discipline. Throughout history, a number of great minds have championed an altruistic message, and aroused in a large number of people a sense of responsibility toward others. Mahatma Gandhi, Martin Luther King, the Dalai Lama, and Mother Teresa are a few inspiring examples.

T: Yes, although we mustn't forget that the lives of both Gandhi and Martin Luther King were cut tragically short by the bullets of assassins. But I would agree that their messages of nonviolence did more for their causes than violence would have done, and humanity at large has been profoundly affected.

Given the Buddhist belief in such action, are Buddhist communities as actively involved in humanitarian work as, for example, Christian societies are?

M: Much remains to be done in this field. Just look at the Tibetan community in exile. Now that it has overcome the traumas of its exodus, the Dalai Lama continually reminds the community that it must start consecrating itself to humanitarian work. He encourages the Tibetans to build schools and clinics for the underprivileged inhabitants of the countries that have welcomed them, saying that Buddhists should imitate the example of their Christian brothers and sisters and act with the same sort of selfless and unflagging determination to relieve suffering.

Of course, much has been accomplished that must not be overlooked. In India, Dr. Ambedkar, an ally of Gandhi and architect of the Indian constitution, who came from the caste of untouchables, converted himself to Buddhism. Not only did he reintroduce Buddhism to India, its country of origin, but he spent his life trying to improve the lives of the untouchables. In Thailand, Buddhist monasteries are the main centers for treating AIDS sufferers and reforming drug addicts. In 1991, Abbot Prajak Kutajitto put up a vigorous resistance, along with the inhabitants of a small village called Pakham, against some large financial companies that wanted to destroy a vast forest. For this, he has endured reprisals and imprisonment. In Burma, Aung San Suu Kyi, who won the Nobel Peace Prize, has continued to lead a nonviolent resistance against the military junta that has been in power since 1988. In New York, Bernard Glasmann has set up a worldwide Buddhist network of assistance for the homeless and environmental protection.[4] He thinks that social work and spiritual practice are one and the same. In the monastery where I live, we have recently opened a large clinic to help the destitute population in the area.

The notion of mutual assistance is thus very much alive in Buddhist societies, even at a governmental level. But we must never lose sight of the necessity of achieving inner discipline and the importance of personal transformation.

T: This emphasis on the inner life leads many to think of Buddhism as a religion. Is Buddhism competitive with the other religions of the world, or is it rather complementary to them?

M: As the Dalai Lama has said, over half of the world's population are nonbelievers. Many say that they are Catholics, Protestants, Hindus, or Muslims because they were brought up in those traditions. But when faced by daily crises, or the great decisions in their lives, they don't really take the teachings of their religion into account. Only a minority now think and act in accordance with their faith.

So we must distinguish between spirituality in general terms, which aims to make us better people, and religion. Adopting a religion remains optional, but becoming a better human being is essential. That's why the Dalai Lama speaks of "secular spirituality," even though this concept may shock representatives of other religions. To his mind, we can't exclude half of humanity from spirituality for the simple reason that they aren't religious believers. From the day we were born until we die, we need both to give and receive tenderness and goodness. When the great religions and spiritual traditions are practiced correctly, then they can help people to develop love, compassion, patience, and tolerance. But those who don't feel any religious inclinations shouldn't be excluded from this process.

T: How do you see the possible development of this secular spirituality in the West?

M: Family upbringing and education in general must put the emphasis back on human, ethical values that assist inner transformation. Parents and teachers, who are often as lost as their charges in this field, think that spirituality is a private affair and doesn't concern education. But it seems to me that schools should offer children the possibility to discover the world's great spiritual traditions—and not just their histories, but also the essence of their teachings and ethics. I think that seeing secularity as a total absence of spiritual education is an impoverishment and a block to intellectual freedom. Since many young people have never been confronted with ideas that might inspire them, they think that life is meaningless.

T: But does this justify turning schools into religious "supermarkets"? I agree that there's a danger of losing our frames of reference, but then there's also a real danger of a rise in fundamentalism. We must be very careful about mixing education and religion.

M: I don't mean that we should put various dogmas on offer, and try to pick up as many believers as possible. On the contrary, the approach I would advocate would be free of any partisanship and would give young people a complete overview of what the great spiritual traditions have to offer. Such an education would highlight points of agreement as well as differences, and would avoid assigning greater or lesser importance to any set of beliefs, including agnosticism. The problem we have now is an absence of ways to give meaning to our lives.

T: This idea of lay spirituality does seem to me to address some of Western societies' preoccupations. Personally, it suits me fine. I try to live according to certain principles I acquired during my Buddhist upbringing, while continuing to do my job as an astrophysicist. Your case, on the other hand, is more extreme. You left Western society and scientific research and became a monk in a Tibetan monastery in Nepal. Obviously, we can't all follow the same path you have.

M: That's what I was trying to say when I pointed out that spirituality is for those who live in the world just as much as for those who have chosen a contemplative life. If it were restricted to monks and nuns, then it wouldn't be half of humanity, but 99.99 percent, that would be excluded! Spirituality begins with work on our minds, which we are all capable of. However, if we acquire only theoretical knowledge, no matter how complete it may be, we run the risk of becoming one of those people who never get it wrong, except when it comes to the essential!

THE MONK'S CONCLUSION

Many people think that any attempt to bring science and spirituality together is doomed to failure. Some consider spirituality to be hocus-pocus, others that science is too materialistic, and others still that the two fields are totally incompatible. This refusal to find any common points between them comes down to saying that there are impassable barriers between knowledge and experience, subject and object, or matter and consciousness. Such duality is stubbornly persistent and leads to the setting up of unnecessary borders. For the contemplative approach does not entail swimming upstream against true science, but instead means allotting priorities to different fields of knowledge and to the means used to gain access to them.

What matters most in life is not the quantity of information that we can acquire, but the answers to questions such as, why are we alive? why do we die? why do we suffer? why are we happy? why do we love? why do we hate? This should lead us to wonder if the object of our research is able to come up with any answers, and if it really merits all the time we consecrate to it.

In science, such questions concern two fields: fundamental research and its applications. Fundamental research's aim is to describe and explain nature in an "objective" way. This may well be an excellent project, but our curiosity about the chemical makeup of the stars, or the classification of insects, must surely take second place to basic questions concerning our existence. If we think about the moments that give most meaning to our lives, we generally mention love, friendship, tenderness, joy, the beauty of a natural landscape, inner peace, or altruism. Science that simply focuses on outer phenomena has very little to do with any of these.

As for science's applications, they generally concern our health, life expectancy, freedom of action, and comfort. We now live much longer. The general quality of health care is constantly improving, despite shocking disparities ($2,765 per year and per American for public and private care, but only three dollars for each Vietnamese). If our material freedom, our physical comfort, and our ability to act on the world continue to grow, several other aspects of our existence have worsened. We have polluted just about everything that it is possible to pollute, and new disasters confront our planet and its inhabitants. Does the dogma of continual technological and economic growth really deserve the pedestal we have placed it on? We need to focus on how this dogma causes confusion between what is possible and what is desirable.

Only after science had given up all hope of being able to know everything about everything did it make its greatest leaps forward. By concentrating on natural phenomena, it has found efficient ways to discover, measure, and describe them, and then to act on them. There is now so much knowledge of this sort that its vastness sometimes makes us forget that science is incapable of answering basic questions about existence. But this must not be seen as a failing, given that science has clearly marked out its field of investigation and what it can do. Making us happy or establishing peace around us has never been its objective.

Science studies "observables," and physical theories explain more or less accurately the phenomena under observation. Niels Bohr wrote, "The task of science is both to extend the range of our experience and to reduce it to order,"[1] and "In our description of nature, the purpose is not to disclose the real essence of phenomena, but only to track down, so far as possible, relations between the manifold aspects of our experience."[2] Thus there is no way that science will ever understand all of nature, as was once hoped. The scientific method generally leads only to an interpretative system that allows us to describe phenomena and predict their experimental behavior. Science has finally come up against some obstacles that show that the nature of reality is not what was once thought. These very obstacles, which have been revealed by quantum mechanics and relativity, have led science into starting a dialogue with Buddhism. In its examination of the ultimate nature of phenomena and of consciousness as a step toward enlightenment, Buddhism can offer some answers to the scientists' dilemma of trying to reconcile the apparent reality of the macrocosm with the disappearance of solid reality as soon as we enter the world of particles. But it goes much further, because it translates its conclusions into a pragmatic attitude to life.

As for technology, it sees science as a way to use the world and to carry on its dream of one day mastering it. Fundamental science is theoretical knowledge, while technology is utilitarian knowledge and contemplative science is liberating knowledge. They can thus complete each other without any conflict.

To my mind, the most fascinating part of this confrontation between the natural sciences and Buddhism is in the analysis of the ultimate nature of things. I have learned a lot from our conversations. They have forced me to confront new questions concerning our two disciplines—particularly when it comes to the nature of consciousness and the interdependence of phenomena, which lies at the heart of both modern physics and Buddhist teachings. The nature of consciousness remains a fascinating subject. Can it be totally reduced to the brain? Is it a phenomenon that emerges from matter? Can it—as Buddhism thinks—only be born from preceding instants of consciousness and continue without a physical framework? Buddhist contemplatives speak of different levels of consciousness, which they have defined according to genuine introspective experiments. Their method deserves to be studied by researchers who base their work on science's empirical approach. Until recently, the lack of contemplative experience by most of the scientists who have investigated the workings of the mind has led nowhere in understanding the nature of consciousness. From a Buddhist perspective, it seems much more reliable and informative to train the mind to investigate itself, since it thus has direct access to mental events and to its ultimate nature, than to monitor from outside the corresponding activities of the brain. One would thus avoid falling onto the sterile ground of "eliminative materialists" for whom subjectively experienced mental states must be regarded as nonexistent on the basis that the descriptions of such states are irreducible to the language of neuroscience. One could, on the other hand, envisage a thrilling research project bringing together neurobiologists and contemplatives.

What can be said of Thuan's wager on the existence of an organizing principle in the universe? Any approach to the question of origins forces us to adopt a metaphysical position. As François Jacob said, "One field must be totally excluded from scientific enquiry: the origin of the world."[3] But a metaphysical position is not always the same as a wager, and Buddhism sees no need for gambling. In its opinion, the only metaphysical approach to the question of a beginning that stands up to analysis is the absence of any beginning. Any other possibility inevitably leads to a causeless cause, something immutable that changes itself, or nothing becoming something. If we adopt this finalistic wager, then solutions must be found to this sort of contradiction.

As for Leibniz's question—"Why is there something, rather than nothing?"—which Thuan returns to, it is meaningful only from a realist/materialist point of view. It presupposes that phenomena have an intrinsic reality. Positing an organizing principle gets us nowhere. The question then becomes "Why is there an organizing principle, rather than nothing?" In Buddhist terms, it could be rephrased as "Why is there a manifestation of phenomena, rather than nothing?" Answer: "Because everything is empty, everything can appear." For Buddhism, there has never been a solid reality with an intrinsic existence. Enlightenment simply consists in awakening from a dream of ignorance that attributes this intrinsic reality to objects.

Thuan remarks that "according to the latest astronomical observations, the universe does not seem to have enough matter to make it stop and reverse its outward motion. Our present state of knowledge thus seems to exclude the idea of a cyclical universe." But this question is far from being closed. Scientists are constantly coming up with new ideas about it. A popular French science magazine, *Science et Vie,* devoted its January 2000 issue to the "pre–Big Bang." In it, eminent scientists give us their revolutionary visions of the universe and cosmology. Andrei Linde speaks of a cascading universe, in which Big Bangs occur at each moment. Martin Rees evokes a "multiverse," made up of numerous universes that are constantly and endlessly regenerated. As for Gabriele Vaneziano, the Big Bang was not the beginning of the universe, but just one major event in its history. The fact that these scenarios are so different, contradictory, and mutually incompatible (and there will certainly be more to come) shows just how precarious any cosmological theory is when it tries to deal with the beginning of the universe.

In the Buddhist viewpoint, this Western obsession—in religion, philosophy, and science—with a beginning derives from a stubborn belief in the reality of phenomena: objects really "exist" as we see them, and so must have a beginning. This approach forces scientists into performing complicated juggling acts when trying to reconcile the results of quantum mechanics with a reassuring vision of the world, thus preserving us from having to put our ordinary perception of things into question. The efforts made by physicists to hang on to certain classical representations (such as in the inherent existence of material bodies with intrinsic properties) come into inevitable conflict with phenomena that they themselves have discovered. Such difficulties do not arise only from the inertia that always results from belonging to a particular scientific school; they come from a far deeper resistance that rises up whenever doubts are raised about the reality of phenomena and of the subject observing them. Michel Bitbol has observed that the philosophical debate concerning modern physics seems to be dominated by the following maxim: "Whenever a realist interpretation is available in theoretical physics—whatever happens, adopt it at once."[4] Yet if physicists drew the logical conclusions from quantum mechanics, then they could easily transform their worldview.

When the Nobel physics laureate Steven Weinberg remarks that "it takes religion for good people to do evil," an equally dogmatic response would be, "Only spirituality can make evil beings do good." Apart from the atrocities that are sometimes committed in the name of science (some examples of which were given in chapter 1), it can also be stated that only science gives an excuse for normal people to do wrong under the cloak of respectability. But that would be to adopt a position as extreme as Weinberg's. It would be more accurate to say that the value of any activity, even such respectable ones as science and religion, depends entirely on our motivations.

I do not think that what I call "contemplative science" is mainly a question of intuition. Such a term seems too vague to describe the direct experience of meditation, which would be valueless if it were not based on valid cognition. On a certain level of contemplation, reason must be transcended, but this does not mean that meditation then flies in the face of reason once this level has been passed. It quite simply surpasses its limitations. Thuan compared this process to Gödel's theorem of incompleteness. Buddhist philosophy and contemplation do not set out to construct a grandiose theoretical edifice. But what they do insist on are tangible results in terms of inner transformation. Khyentse Rinpoche gave the following advice on this subject:

> *The sign of wisdom is self-control, and the sign that we have matured in our spiritual experience is a lack of conflicting emotions. This means that when we have become wise and knowledgeable, we should have become serene, peaceful, and disciplined to the same degree—and not negligent, arrogant, or puffed up with pride. Constantly check that you reuse spiritual practices to tame your negative emotions. But if a given practice has the opposite effect and increases your egoism, your confusion and your negative thoughts, you would do better to abandon it, because it is not meant for you.*

If we must trust the instructions of accomplished teachers, whose experience is greater than ours—just as we listen to the advice of a well-traveled sailor—we must not accept truths just because they have been pronounced by someone we respect. The value of the Buddha's words lies in the fact that we all can check them out for ourselves. François Jacob wrote, "The danger for scientists [and, I would add, for contemplatives] is not to measure the limits of their science, and thus of their knowledge. This leads them to mix what they believe and what they know. Above all, it creates the certitude of being right."[5]

In the natural sciences, successive revolutions have shown that we can never be sure about being "definitely right." So dare I say that inner enlightenment provides a different sort of certitude about the ultimate nature of the mind, the mechanisms of happiness and suffering, and the reality of phenomena? This certitude comes from an inner discovery that is confirmed at each moment of our existence. It appears as an immutable understanding of the true nature of things, and manifests itself in the human qualities that we all want to have.

For, in the end, if we do not have such qualities, what is the point of erecting endless intellectual constructs, like castles built on a frozen lake? Just as they will vanish into the waters when spring comes, purely conceptual meditation, which does not cause profound changes in our being, will not stand up to the trials of existence.

According to Buddhism, the understanding of emptiness leads to boundless love and compassion. Shabkar, the Tibetan hermit, wrote:

With compassion, one has all the teachings,
Without compassion, one has none of them.
Even those who meditate on emptiness
Need compassion as its essence.[6]

To use a metaphor found in Buddhist texts, only the heat of that compassion united with wisdom can melt the ore in our minds, so as to liberate the gold of our fundamental nature.

THE SCIENTIST'S CONCLUSION

At the close of our conversations, I must say that my admiration for how Buddhism analyzes the world of phenomena has grown considerably. At the beginning of this project, I was rather skeptical. I was familiar with, and appreciated, Buddhism's practical side, which provides a guide for self-knowledge, spiritual progress, and becoming a better human being. So far as I knew, Buddhism was primarily a pathway leading to enlightenment, a contemplative approach with an essentially inward gaze.

I knew that science and Buddhism had radically different methods for investigating reality. In science, intellect and reason have the leading roles. By dividing, categorizing, analyzing, comparing, and measuring, scientists express natural laws in the highly abstract language of mathematics. Intuition is not excluded from science, but it can play a part only if it can be formulated in a coherent mathematical structure. By contrast, it is intuition—or inner experience—that has the leading role in the contemplative approach, which refuses to break up reality, but instead aims to understand it in its entirety. Buddhism has no use for measuring apparatus, and does not rely on the sort of sophisticated observations that form the basis of experimental science. Its statements are more qualitative than quantitative. So I was far from sure that there would be any point in confronting science with Buddhism. I was afraid that Buddhism would have very little to say about the nature of the world of phenomena, because this is not its main interest, whereas such preoccupations lie at the heart of science. If that had been the case, then we would probably have ended up with two parallel discourses without ever finding common ground.

But as our conversations progressed, I soon realized that my fears were groundless. Not only has Buddhism thought about the nature of the world, but it has done so in a deep and original way. Its purpose is not to find out about the world of phenomena for its own sake, but because it is by understanding the true nature of the physical world—emptiness, interdependence—that we can clear away the mists of ignorance open the way to enlightenment. Our discussions were mutually enriching. They led to new questions, original viewpoints, and unexpected syntheses that required further study and clarification, and still do so.

These conversations form part of an ongoing dialogue between science and Buddhism. The most important thing that they taught me was that there is a definite convergence and resonance between the Buddhist and scientific visions of reality. Some of Buddhism's views on the world of phenomena are strikingly similar to the underlying notions of modern physics—in particular, its two main grand theories: quantum mechanics, which is the physics of the infinitely small; and relativity, the physics of the infinitely large. Even though Buddhism and science have radically different ways of investigating the nature of reality, this does not lead to an insuperable opposition, but rather to a harmonious complementarity. That is because both are on quests for the truth, and both use criteria of authenticity, rigor, and logic.

Take, for example, one of Buddhism's central tenets, the "interdependence of phenomena." Nothing exists inherently, or is its own cause. An object can be defined only in terms of other objects. Interdependence is essential to the manifestation of phenomena. The world would not be able to function without it. So a given phenomenon can come about only if it is linked to others. Reality cannot be localized and fragmented, but should be considered as holistic and global.

Several experiments in physics have now imposed this global view on us. In the atomic and subatomic world, EPR-type experiments have shown us that reality is "inseparable." Two light particles that have interacted continue to act as parts of a single reality. However far apart they are, they behave in an instantaneously correlated way, without any exchange of information occurring. As for the macroscopic world, its global nature is shown by Foucault's pendulum, whose behavior does not depend on its local environment, but on the entire universe. What happens on our little planet is determined in the vast immensity of the cosmos.

The concept of interdependence states that things cannot be defined in absolute terms, but only in relation to others. This is, in substance, the same idea as the principle of the relativity of motion in physics, which was first stated by Galileo and then developed to its perfection by Einstein. "Motion is as nothing," Galileo stated. What he meant was that an object's motion could not be defined in absolute terms, but only in relation to the motion of a second object. There is no way for passengers on a train moving at a constant speed, with all of the windows closed, to find out by measurement or experimentation whether the train is moving or at a standstill. It is only by opening a window and looking at the countryside speeding past that the passengers can find out. As long as there is no exterior frame of reference, then motion is equivalent to non-motion. Buddhism says that objects do not exist inherently, but only in relation to others. The relativity principle says that the train's motion exists only in relation to the passing countryside.

Time and space have lost the absolute characteristics that Newton gave them. Einstein showed us that they can be defined only in relative terms depending on the motion of the observers and the intensity of the field of gravity around them. In the vicinity of a "black hole," one second can stretch to eternity. As in Buddhism, relativity teaches us that the idea of a past already gone and a future still to come is mere illusion, given that my future can be someone else's past and a third person's present—it all depends on our relative motions. Times does not pass, it simply is there.

The notion of interdependence leads us directly to the idea of emptiness, which does not mean nothingness, but the absence of inherent existence. Since everything is interdependent, nothing can be self-defining and exist inherently. The idea of intrinsic properties that exist in themselves and by themselves must thus be thrown out. Once again, quantum physics has something strikingly similar to say. According to Bohr and Heisenberg, we can no longer talk about atoms and electrons as real entities with well-defined properties, such as speed and position; we must now consider them as part of a world made up of potentialities and not of objects and facts. The very nature of matter and light becomes subject to interdependent relationships. It is no longer intrinsic, but can change because of an interaction between the observer and the object under observation. Such a nature is not unique anymore, but dual and complementary. The phenomenon that we call a "particle" becomes a wave when we are not observing it. But as soon as a measurement or observation is made, it starts looking like a particle again. To speak of a particle's intrinsic reality, or the reality it has when unobserved, would be meaningless because we could never apprehend it. As in the Buddhist notion of *samskara,* or "event," quantum mechanics has radically relativized our conception of an object, by making it subordinate to a measurement or, in other words, an event. What is more, quantum uncertainty places a stringent limit on how accurately we can measure reality. There will always be a degree of uncertainty about either the position or the speed of a particle. Matter has lost its substance.

The Buddhist notion of interdependence is synonymous with emptiness, which is in turn synonymous with impermanence. The world is like a vast stream of events and dynamic currents that are all interconnected and constantly interacting. This concept of perpetual, omnipresent change chimes with modern cosmology. Aristotle's immutable heavens and Newton's static universe are no more. Everything is moving, changing, and impermanent, from the tiniest atom to the entire universe, including the galaxies, stars, and mankind.

The universe is expanding because of the impulse it received from its primordial explosion. This dynamic nature is described by the equations of the Theory of Relativity. With the Big Bang theory, the universe has acquired a history. It has a beginning, a past, present, and future. One day it will die in an infernal conflagration or else an icy freeze. All of the universe's structures—planets, stars, galaxies, and galaxy clusters—are in perpetual motion and take part in an immense cosmic ballet: they rotate about their own axes, orbit, fall toward or move apart from one another. They, too, have a history. They are born, reach maturity, then die. Stars have life cycles that span millions or even billions of years.

The same goes for the atomic and subatomic world. There, too, everything is impermanent. Particles can change their nature: a quark can change its family, or "flavor," a proton can become a neutron and emit a positron and a neutrino. Matter and antimatter annihilate each other to become pure energy. The energy of a particle's motion can be transformed into another particle, or vice versa. In other words, an object's property can become an object. Because of the quantum uncertainty of energy, the space around us is filled with an unimaginable number of "virtual" particles with fleeting, ghostlike existences. Constantly appearing and disappearing, they are a perfect illustration of impermanence, with their infinitely short life cycles.

So reality can be perceived in various ways, and different approaches—one turned inward and the other outward—can lead to the same truths. Buddhism will surely not find such agreement surprising. Since the world of phenomena can be observed only through the filter of consciousness, and given that consciousness itself is interdependent on the exterior world, the fundamental nature of phenomena cannot be alien to the Buddha's enlightened mind.

However, I do have reservations about how Buddhism deals with the "anthropic principle," according to which the universe's physical constants and initial conditions were exquisitely fine-tuned to allow for the emergence of life and consciousness. To account for that fine-tuning, I made a Pascalian wager on the existence of a creative principle. This principle—which I see in the same terms as Spinoza or Einstein—manifests itself in natural laws and is the reason why the world is rational and intelligible. This position is contrary to the Buddhist approach, which refuses to admit a creative principle (or a watchmaker God). It considers that the universe doesn't need tuning for consciousness to exist. Since both coexist fundamentally, they cannot exclude each other. Once more, interdependence offers a solution. While I admit that this might explain the fine-tuning of the universe, it seems far less clear to me that it answers Leibniz's existential question: "Why is there something, rather than nothing?" I would add, "Why are the natural laws as they are and not different?" For example, it would be quite easy to imagine us living in a universe governed only by Newton's laws. But this isn't the case. For the laws of quantum mechanics and relativity describe the known universe.

The Buddhist view also raises other questions. If there is no Creator, the universe cannot have been created. So there is neither a beginning nor an end. The only sort of universe that would be compatible with this idea is a cyclical one, with an endless series of Big Bangs and Big Crunches. But the scenario of the universe one day collapsing into itself in a Big Crunch is far from being proven scientifically. It all depends on the total amount of dark matter in the universe, and this is as yet unknown. According to the latest astronomical observations, the universe does not seem to have enough dark matter to stop and then reverse its expansion. Our present state of knowledge thus seems to exclude the idea of a cyclical universe. As for streams of consciousness that have coexisted with the universe since the first fractions of a second after the Big Bang, science is far from being able to examine this question. Some neurobiologists think that there is no need for a consciousness continuum that coexists with matter, and that the former can emerge from the latter, once a certain complexity threshold has been passed.

Made of stardust, we share the same cosmic history as the lions on the savannas and the lavenders in the fields. We are all connected through time and space, and thus interdependent. Just breathing links us to the rest of humanity—the billions of oxygen molecules that we inhale with each breath have at some time or other been inside the lungs of each of the 50 billion individuals that have lived on Earth. This cosmic and planetary perspective emphasizes not only our interdependence, but also how vulnerable our planet is, and how isolated we are among the stars. The environmental problems that endanger our haven in the cosmic immensity transcend the barriers of race, culture, or religion. Industrial toxins, radioactive waste, and the greenhouse gases responsible for global warming ignore national borders. These and other problems—poverty, war, famine—that threaten humanity can be solved if we realize that we are interdependent and that our interests and happiness are inextricably bound up with those of others. In other words, we should let compassion guide us and so develop our sense of what the Dalai Lama so rightly calls our "universal responsibility."

Science must resume its proper place in human culture, from which it has drawn somewhat away because its vision has been too fragmentary, mechanistic, and reductionist. But this is no longer true, as these conversations have vividly shown. Certainly, science will have an increasingly large impact on our lives. When faced with ethical or moral problems, which, as in genetics, are becoming ever more pressing, science needs the help of spirituality in order not to forget our humanity. The aim of science is to understand the world of phenomena. Its technical applications can have a good or bad effect on our physical existence. Spirituality, however, aims to improve our inner selves so that we can improve everybody's existence. Some people, like the Nobel prize–winning physicist Steven Weinberg, take a dim view of spirituality. In a typically provocative vein, he writes: "With or without religion, good people can behave well and bad people can do evil; but for good people to do evil—that takes religion. . . . I am all in favor of a dialogue between science and religion, but not a constructive dialogue. One of the great achievements of science has been, if not to make it impossible for intelligent people to be religious, then at least to make it possible for them not to be religious. We should not retreat from this accomplishment."[1] He then goes on to cite many of religion's evils: the Crusades, the pogroms, the Jihad and other religious wars, and even slavery. I think he is wrong. First, he forgets to mention all the evil science can do when in the wrong hands: Hiroshima and Nagasaki, global warming, the hole in the ozone layer,[2] the "research" carried out by Nazi doctors, and so on. Examples are not lacking. Furthermore, the religion Weinberg is talking about (I prefer the word "spirituality") is not the "true" one, but one of its distorted versions.

Those who fight in religious wars cannot possibly be moved by the compassion for others that lies at the root of all true religions. In contrast to Weinberg's antireligious stance, I prefer by far Einstein's cosmic vision, with which I am much more in harmony: "The religion of the future will be a cosmic religion. It will have to transcend a personal God and avoid dogma and theology. Encompassing both the natural and the spiritual, it will have to be based on a religious sense arising from the experience of all things, natural and spiritual, considered as a meaningful unity." These words are very close in spirit to the substance of our conversations. Einstein continued: "Buddhism answers this description. . . . If there is any religion that could respond to the needs of modern science, it would be Buddhism."[3] No one could put it better.

Science can operate without spirituality. Spirituality can exist without science. But man, to be complete, needs both.

NOTES

INTRODUCTION

1 Among the numerous participants in such seminars can be mentioned Richard J. Davidson, Anne Harrington, Jerome Engel, Paul Ekman, Robert Livingstone, Eliot Sober, and Francisco Varela for the cognitive sciences and biology; David Finkelstein, Arthur Zajonc, and Anton Zeilinger for physics; and Owen Flannagan, Daniel Goleman, Charles Taylor, and Lee Yearly for philosophy.

2 B. Alan Wallace, *The Taboo of Subjectivity: Toward a New Science of Consciousness* (New York: Oxford University Press, 2000), 11.

3 See Jeremy Hayward and Francisco Varela, eds., *Gentle Bridges: Conversations with the Dalai Lama on the Science of Mind* (Boston: Shambhala, 1992); Daniel Goleman, ed., *Healing Emotions: Conversation with the Dalai Lama on Mindfulness, Emotions, and Health* (Boston: Shambhala, 1997); Francisco J. Varela, ed., *Sleeping, Dreaming and Dying: An Exploration of Consciousness with the Dalai Lama* (Boston: Wisdom Publications, 1997). See R. J. Davidson and A. Harrington, eds., *Science and Compassion: Dialogues Between Biobehavioral Scientists and the Dalai Lama* (New York: Oxford University Press, in press). See also Zara Houshmand, Robert B. Livingstone, and B. Alan Wallace, eds., *Consciousness at the Crossroads: Conversations with the Dalai Lama on Brain Science and Buddhism* (Ithaca, N.Y.: Snow Lion, 1999). Dalai Lama et al., *Mind Science: An East-West Dialogue,* edited by Daniel Goleman and Robert A. F. Thurman (Boston: Wisdom Publications, 1991); Paul Ekman, Richard Davidson, Matthieu Ricard, and B. Alan Wallace, *Buddhist and Psychological Perspectives on Emotions and Well-Being* (in press).

4 Francisco Varela, E. Thomson, and E. Rosch, *The Embodied Mind* (Boston: MIT Press, 1991).

5 B. Alan Wallace, *Choosing Reality* (Ithaca, N.Y.: Snow Lion, 1996), and *The Taboo of Subjectivity* (see above).

6 Werner Heisenberg, "Wolfgang Pauli's Philosophical Outlook," chap. 3 in *Across the Frontiers* (New York: Harper & Row, 1974).

7 See Trinh Xuan Thuan, *The Birth of the Universe: The Big Bang and After* (New York: Abrams, 1993); *The Secret Melody* (New York: Oxford University Press, 1995); *Chaos and Harmony* (New York: Oxford University Press, 2000).

CHAPTER 1. AT THE CROSSROADS

1 Dilgo Khyentse, *Le Trésor du Coeur* (Paris: Editions de Seuil, Points Sagesses, 1994).

2 For a presentation of the Buddhist concept of emptiness, see *Wisdom: Two Buddhist Commentaries* (Paris: Editions Padmakara, 1993).
3 For a good edition of the treatise, see *The Perfection of Wisdom,* R. C. Jamieson, Viking Press, 2000.
4 Figures from the report of the "United Nations' Development Program" on the state of the world in 1998.

CHAPTER 2. TO BE AND NOT TO BE

1 The theory of relativity published by Einstein in 1915, which describes the effects of gravity on motion, time, and space, is called "General" to distinguish it from the Special Theory of Relativity that Einstein devised ten years earlier, and which does not take the effects of gravity into account.
2 For a detailed account of the Big Bang theory, see Trinh Xuan Thuan, *The Secret Melody* (New York: Oxford University Press, 1995).
3 The weak and strong nuclear forces, the electromagnetic force, and gravity. All physical phenomena can be explained as interactions mediated by these four fundamental forces.
4 At Planck time, the universe was so compressed and had such high density that gravity, which normally has a very slight effect at a subatomic level, was as important as the weak and strong nuclear forces and the electromagnetic force.
5 For a clear, detailed description of the theory of superstrings, see Brian Greene, *The Elegant Universe* (New York: W. W. Norton, 1999).
6 The laws of physics have been tested only for times greater than about a millionth of a millionth of a second (10^{-12}) after the Big Bang. Before then, the energy of the universe's elementary particles was too great for our particle accelerators to be able to reproduce it, so that the physics of these extreme conditions cannot yet be studied experimentally.
7 See Shantideva, *Bodhicaryvatara: The Way of the Bodhisattva,* vol. 9, vv. 147–48 (Boston: Shambhala, 1997).

CHAPTER 3. IN SEARCH OF THE GREAT WATCHMAKER

1 Blaise Pascal, *Pensées,* translated by A. J. Krailsheimer (New York: Penguin Books, 1995).
2 Jacques Monod, *Chance and Necessity* (New York: Alfred A. Knopf, 1971). Steven Weinberg, *The First Three Minutes* (New York: Basic Books, 1977).
3 Freeman Dyson, *Disturbing the Universe* (New York: Harper & Row, 1979).
4 According to the superstrings expert Brian Greene, who was later questioned by Matthieu: "The answer lies in cosmology. In the universe's initial moments, its energy was in the form of vibrating strings. So why did some strings start vibrating in one way, and another lot in a different way? We think that there was competition between entropy (the physical quantity that measures 'disorder'; the universe seems to 'look for' maximum disorder) and energy (the universe looks for lower and lower states of energy, like a ball rolling downhill), which determined how many strings vibrated in each possible way."

5 The strong and weak versions of the anthropic principle are stated by Brandon Carter in *Confrontation of Cosmological Theories with Observation,* edited by M. S. Longair (Dordrecht, Netherlands: Reidel, 1974), 291.
6 True chance cannot be demonstrated, however, because the consequences of indeterminism and deterministic chaos are indistinguishable.
7 A black hole is a singularity in space caused by the gravitational collapse of a massive star at the end of its life. The hole is "black" because its gravity is so great that no light can escape from it.
8 François Jacob, *Of Flies, Mice and Men* (Boston: Harvard University Press, 1998).
9 See Trinh Xuan Thuan, *Chaos and Harmony,* chaps. 2–6 (New York: Oxford University Press, 2000).
10 *The Quotable Einstein,* edited by Alice Calaprice (Princeton: Princeton University Press, 1996), 151.
11 Ibid., 147.
12 Bertrand Russell, *Why I Am Not a Christian* (New York: Simon & Schuster, 1967), 7.
13 In his "Commentary on Logic" *(Pramana-vartika),* Dharmakirti writes, "What can prevent the effect of a complete cause?"
14 Dalai Lama, *The Good Heart: A Buddhist Perspective on the Teachings of Jesus* (Boston: Wisdom Publications, 1996), 73.
15 Ibid., 73.
16 Ibid., 74.
17 Ibid., 74.

CHAPTER 4. THE UNIVERSE IN A GRAIN OF SAND

1 For a detailed treatment of interdependence and mutual causality, see Joanna Macy, *Mutual Casuality in Buddhism and General Systems Theory* (Albany: State University of New York Press, 1991).
2 Glenn H. Mullin, *Selected Works of the Dalai Lama,* vol. 7 (Ithaca, N.Y.: Snow Lion, 1985), 118.
3 Fabrice Midal, *Les mythes et dieux tibétans* (Paris: Éditions de Seuil, 2000).
4 E. Schrödinger, "Mind and Matter," in *What Is Life?* (New York: Cambridge University Press, 1983), 127.
5 This can be compared with Kant, who distinguished between the conventional "objective" existence and autonomous "per se" existence.
6 Or rather, to avoid particle ontology, which is criticized below, we can say that the behavior of phenomenon *a* is always exactly correlated with that of *b.* According to the French philosopher of science Michel Bitbol, quantum physics has nothing to say about reality; it quite simply predicts phenomena. Then it is up to us to try to interpret these predictions by using metaphysical images. Here we have presented the opinion of the majority of physicists. A few isolated scientists, such as David Bohm, have come up with different interpretations of quantum physics, which lead to other interpretations of the EPR experiment.
7 Bernard d'Espagnat, *Veiled Reality* (Reading, Pa.: Addison-Wesley, 1995).
8 Werner Heisenberg, *Physics and Philosophy* (London: Penguin Books, 1990).

9 William Blake, *Auguries of Innocence,* in *Complete Writings,* edited by Geoffrey Keynes (New York: Oxford University Press, 1985), 431.
10 *Avamtasaka Sutra: The Flower Ornament Scripture,* translated by Thomas Cleary (Boston and London: Shambhala, 1993), 959.
11 Asvaghosa, *The Awakening of Faith,* translated by Hakeda (New York: Columbia University Press), 55.
12 *Wisdom: Two Buddhist Commentaries* (Paris: Éditions Padmakara, 1993), 114.
13 Ibid., 114.
14 Ibid., 241.
15 Nagarjuna, "The Fundamental Treatise on the Middle Way," in ibid., 240.
16 Ibid., 242–43.

CHAPTER 5. MIRAGES OF REALITY

1 Sir James Jean wrote: "We can never understand what events are, but must limit ourselves to describing patterns of events in mathematical terms; no other aim is possible. Physicists who are trying to understand nature may work in many different fields and by many different methods; one may dig, one may sow, one may reap. But the final result will always be a sheaf itself. . . . [Thus] our studies can never put us into contact with reality." *Physics and Philosophy* (Cambridge, England: Cambridge University Press, 1931), 15–17.
2 Shantideva, *Bodhicaryvatara: The Way of the Bodhisattva,* vol. 9 (Boston: Shambhala, 1997), 34.
3 In counterpoint to this analysis of Buddhism and quantum physics, Laurent Nottale wrote: "The concept of a particle would thus no longer concern an object that 'had' a mass, a spin, or a charge, but would correspond to the fractal geodesics of a nondifferential space-time. Geodesics such as mass, spin, and charge would thus be common geometric properties." *La Relativité dans tous ses états* (Paris: Hachette, 1998), 238.
4 This argument, devised by Chandrakirti (eighth century C.E.), the great Indian commentator of Nagarjuna, is generally applied to a chariot. The seven reasons why a "chariot" has no inherent existence can be summarized as follows:

1. A chariot is *not intrinsically the same* as its parts (wheels, axles, etc.), for they are multiple, and the entity of a chariot would thus become multiple. If one insists that the chariot really is "one" entity, then all of its parts must be a single entity. Thus, absurdly, the agent (the moving chariot) and that which draws it along (its parts) would be one.
2. A chariot is *not intrinisically different* from its parts, for if it were, then it would be an entity totally distinct from its parts. But ontologically, independent and simultaneous phenomena cannot act on one another, and so cannot be connected by a causal chain. The chariot should then be perceived as being separate from its parts, which is not the case.
3. The parts of a chariot *do not depend intrinsically* on the entire chariot, for if they did, then the parts and the whole of the chariot would have to be

intrinsically "different," which returns us to the previous point.

4. For the same reasons, a chariot *does not intrinsically depend* on its parts.
5. A chariot *does not possess* its parts, as a farmer owns a cow or a man his body. For that to be true, the chariot would have to be either intrinsically distinct, or indistinguishable from its parts. Both of these possibilities have already been refuted.
6. The chariot entity is *not a simple composite* of its parts: (a) the form of its parts cannot be a chariot; and (b) the form of the composite made up by the parts cannot be a chariot, because the forms of these parts remain unchanged, i.e., they are a chariot neither before nor after coming together.
7. The *form of the composite* is not a chariot, because the composite formed by the parts is not an entity with a distinct existence. There is no composite of the parts different from the parts themselves, otherwise we could perceive the composite without perceiving the parts. As we have seen, the composite cannot be identical to its parts, for if it were, either the "composite" entity would be multiple, or the parts would be a single object. To sum up, the form of the composite exists only through a *conceptual imputation.*

5 B. Alan Wallace, *Choosing Reality* (Ithaca, N.Y.: Snow Lion, 1996), 120.

6 When trying to understand Bohr's principle of complementarity, we must not overlook his essential idea: "The impossibility of any separation between the behavior of atomic objects and the interaction with measuring instruments used to define the conditions in which the phenomena take place." Niels Bohr, *Atomic Physics and Human Knowledge* (New York: John Wiley, 1958).

7 Werner Heisenberg, *Physics and Beyond: Encounters and Conversations* (Harper & Row, New York, 1971).

8 ———, *Physics and Philosophy* (London: Penguin Books, 1990), 131–33.

9 Niels Bohr, *Atomic Theory and the Description of Nature* (Woodbridge, Conn.: Ox Bow Press, 1987), 18.

10 François Jacob, *Of Flies, Mice and Men* (Boston: Harvard University Press, 1998).

11 E. Schrödinger, *Science and Humanism* (Cambridge, England: Cambridge University Press, 1951), 47.

12 Heisenberg, *Physics and Philosophy,* 95.

13 Henry Stapp, *S-Matrix Interpretation of Quantum Theory,* Lawrence Berkeley Laboratory reprint, 22 June 1970 (revised edition, *Physical Review,* D3, 1971, 1303).

14 Michel Bitbol, *L'Aveuglante Proximité du Réel* (Paris: Flammarion), 218.14.

15 E. Schrödinger, *Nature and the Greeks* (Cambridge, England: Cambridge University Press, 1954), 85.

16 W. V. Quine, *The Pursuit of Truth* (Boston: Harvard University Press, 1990), 35 (13).

17 Laurent Nottale, *La Relativité dans tous ses états* (Paris: Hachette, 1998), 111.

18 Markus Arndt, Olaf Nairz, Julian Vos-Andreae, Claudia Keller, Gerbrand Van der Zouw, Anton Zeilinger, "Wave-particle duality of C60 molecules," *Nature* 401, no. 6754 (1999): 680–82.

19 Stapp, *S-Matrix Interpretation of Quantum Theory.*

20 According to this viewpoint, space-time is fractal at a microscopic level, and the transition between fractal and nonfractal behavior on higher levels is equivalent to the replacement of quantum mechanics by classic mechanics (see Nottale, *La Relativité,* 177). The laws of nature must remain valid in all coordinate systems, regardless of the level (ibid., 219).

21 Shantarakshita, *Madhyamaka-alankara* (Tibetan *dbu ma rygen,* the Ornament of the Middle Way).

22 Shantideva, *Bodhicaryvatara: The Way of the Boddhisattva,* vol. 9 (Boston: Shambhala, 1997), 34.

23 R. P. Feynman, R. B. Leighton, and M. Sands, *Feynman Lectures in Physics,* vol. 1, chap. 1 (Reading: Addison-Wesley, 1963), 2.

24 Louis Finot, *Milanda-Panha, Les Questions de Milanda,* translated from the Pali, "Connaissances de l'Orient" (Paris: Gallimard, 1992).

25 Bernard Pullman, *The Atom in the History of Human Thought* (New York: Oxford University Press, 1998), 33.

26 This indivisibility is physical and not mathematical.

27 Heisenberg, *Physics and Philosophy,* 133.

28 Bitbol, *L'Aveuglante Proximité du Réel,* 188.

29 Ibid., 195.

CHAPTER 6. LIKE A BOLT FROM THE BLUE

1 A proton is made of two "up" quarks and one "down" quark, while a neutron has one "up" quark and two "down" quarks.

2 Wagarjana, *Fundamental Treatise on the Middle Way,* in *Wisdom: Two Buddhist Commentaries* (Paris: Éditions Padmakara, 1998).

3 See Brian Greene, *The Elegant Universe* (New York: W. W. Norton, 1999).

4 This theory supposes that the empty space between the galaxies left by the expanding universe is filled in by a continuous creation of matter and galaxies. Instead of a single Big Bang, the steady-state universe experiences a series of "Little Bangs."

5 Isaac Newton, *Opticks: Or a Treatise of the Reflections, Inflections and Colours of Light* (Dover Publications, 1952).

6 Dharmakirti, *The Complete Commentary on Authentic Knowledge (Pramana-varttika karika),* in *Le Précieux Ornement de la libération,* 249.

7 Patrul Rinpoche, *The Words of My Perfect Teacher,* translated by the Padmakara Translation Group, chap. 2, revised edition (Boston: Shambhala, 1998).

CHAPTER 7. EACH TO HIS OWN REALITY

1 W. H. Zurek, "Quantum, classical, and decoherence," Los Alamos reprint, 1992, quoted in M. Lockwood, "'Many-minds' interpretations of quantum mechanics," *British Journal for the Philosophy of Science* 47 (1996), 159–88.

2 Werner Heisenberg, *Physics and Philosophy* (London: Penguin Books, 1990), 190.

3 Niels Bohr, *The Philosophical Writings of Niels Bohr*, vol. 2, *Essays 1933–1957 on Atomic Physics and Human Knowledge* (Woodbridge, Conn.: Ox Bow Press, 1987), 20.

4 Jean-Marc Levy-Leblond, "Le temps, des équations aux mots," *Les Espaces*, December 1999.

5 Bernard d'Espagnat, *Veiled Reality* (Reading, Pa.: Addison-Wesley, 1995).

6 W. Heisenberg, *Philosophical Problems of Quantum Physics* (Woodbridge, Conn.: Ox Bow Press, 1979), 11.

7 Khyentse Rinpoche, from an oral commentary upon the "Hundred Pieces of Advice to Padampa Sangye," in Matthieu Ricard, *Journey to Enlightenment* (New York: Aperture, 1997), 45.

8 What the Buddhas perceive is the "infinite purity of phenomena." One speaks of "purity" because the discriminations between pure and impure, beautiful and ugly, break down in the non-dual understanding of a phenomenon's ultimate nature. Our ordinary "impure" world is the product of delusion. Ignorance itself is a transient, unreal veil. When we become fully aware of this, all of our "impure" perceptions vanish.

9 See E. Thompson, A. Palacios, and F. J. Varela, "Ways of coloring: Comparative color vision as a case study for cognitive science," in *Behavioral and Brain Sciences* 15 (1992), 1–74.

10 In John W. Pettit, *The Beacon of Certainty* (Boston: Wisdom Publications, 1999), 365.

11 Those who have reached the first level *(bhumi)* of the bodhisattvas, and have lessened their attachment to the notion of ego and the reality of phenomena, are capable of pure perception, without mental images. For an analysis of Buddhism's theory of perception, see George B. J. Dreyfus, *Recognizing Reality: Dharmakirti's Philosophy and Its Tibetan Interpretations* (Albany: State University of New York Press, 1997).

12 David Bohm, lecture given at UC Berkeley in 1977.

13 Henri Poincaré, *La Valeur de la Science* (Paris: Flammarion, 1990).

14 B. Alan Wallace, *Choosing Reality* (Ithaca, N.Y.: Snow Lion, 1996), 98.

15 This topic was developed in great detail by Lama Mipham (1846–1912) in his work *The Beacon of Certainty*, and in commentaries on the text by various hands. See John W. Petit's translation, cited above, n. 10.

16 Nagarjuna, *Refutation of Objections (Vigrahavyavartani)*, in *Wisdom: Two Buddhist Commentaries* (Paris: Editions Padmakara, 1993).

CHAPTER 8. QUESTIONS OF TIME

1 Saint Augustine, *Confessions*, chap. 9, edited by Gillian Clark (New York: Cambridge University Press, 1995).

2 *The Quotable Einstein*, edited by Alice Calaprice (Princeton: Princeton University Press, 1996), 61.

3 The second has now been defined as the duration of 9,192, 631, 770 vibrations of an atom of cesium 133 between two energy levels.

4 Gampopa, translated in Matthieu Ricard, *Journey to Enlightenment* (New York: Aperture, 1997), 134.

5 The subatomic particle called the *kaon* is the sole exception to this rule. It does show an "arrow of time," but this tiny arrow is of no importance given that the kaon isn't found in the matter we are made of, or in the stars and galaxies, but only in particle accelerators.

6 If we don't observe the phenomenon "particle," then it appears as a wave and can be present everywhere. Schrödinger's wave function, which allows us to calculate the probability of finding a particle at a given point in space, is reversible in time. Hence there is, a priori, no "time's arrow." But this holds only so long as no measurement is made. As soon as there's a measurement, everything works as though the wave function had been reduced to a single point, and the system becomes irreversible in time. To see how this irreversibility comes about, let's do a thought experiment that consists of inverting Schrödinger's wave function in time. Everything is reversible until the moment of measurement. As soon as this happens, the particle must choose among several possible pasts, just as there were several possible futures before the act of measuring, when time was running the other way. Nothing forces a particle to choose its "true" past. Thus there is irreversibility and a sort of quantum "time's arrow."

7 Aristotle, *Physics* IV, 10, edited by David Bostock (New York: Oxford University Press, 1996), 218a.

8 *Nunc fluens facit tempus, nunc stans facit aeternitatem,* Boethius, *De Consolatione,* chap. 5, 6.

CHAPTER 9. CHAOS AND HARMONY

1 Werner Heisenberg, *Physics and Philosophy* (London: Penguin Books, 1990), 149.

2 It must be said that passing through the barrier of the speed of light is forbidden only for phenomena that carry information. Phenomena that carry no information can, in theory, travel faster than light. Let's take the example of a laser beam projected from Earth and played rapidly across the surface of the Moon. Even though the photons in the laser beam don't travel faster than light, the light zone projected onto the moon's surface can seem to travel faster than light because of the angle of rotation of the beam and the great distance between the Earth and the Moon. Astronomers have observed this phenomenon in radio galaxies (celestial bodies that emit most of their energy as radio waves), where so-called superluminal motions, that is to say motions exceeding the speed of light, have been detected. But such phenomena can't transmit any information and so can't enter into a causal relationship.

3 According to Roland Omnès, the collapse of the wave function shouldn't necessarily be seen as a phenomenon, but simply as a mathematical trick. For further information, see his *The Interpretation of Quantum Mechanics* (Princeton, N.J.: Princeton University Press, 1994), 339–40.

4 M. Tegmark and J. A. Wheeler, *Scientific American,* February 2001, 68.

5 Pierre-Simon de Laplace, *A Philosophical Essay on Probabilities* (New York, Dover, 1951), 4.

6 Henri Poincaré, *Science and Method* (Bristol, England: Thoemmes Press, 1996).
7 For further details about chaotic and nonlinear phenomena, see Trinh Xuan Thuan, *Chaos and Harmony,* chap. 3 (New York: Oxford University Press, 2000).
8 Freedom of choice exists on a conscious level and is included in the unlimited network of causes and conditions. This is one of the factors that allow us to escape from super-determinism. This brings to mind Karl Popper's logical argument to show that we can't forecast our own actions. Prediction becomes one of the determinant causes of an action. If I predict that I'm going to collide with a tree in a given location in ten minutes' time, that possibility makes me avoid the place in question and so the prediction doesn't come true. What's more, if certain events are inextricably linked to our actions, given that we can't predict the actions, we can't predict the events, either. "In other words," Michel Bitbol observed (personal communication), "when the person who predicts is inextricably involved in the production of the predicted phenomena, he *can absolutely not* predict them in a strict (nonprobabilistic) sense, whether or not the laws of nature are considered to be deterministic. The fact there is a close codependence between the person who predicts and the predicted phenomenon can thus explain why the strict prediction of phenomena is impossible and, what is more, why it is impossible to choose between 'true chance' and causal links, given that the impossibility of prediction is independent of the sort of law supposedly in action."
9 This is also similar to Heisenberg's approach: "Let us consider a radium atom, which can emit an alpha-particle. The time for the emission of the alpha-particle cannot be predicted. We can only say that in the average the emission will take place in about two thousand years. . . . We know the foregoing event, but not quite accurately. We know the forces in the atomic nucleus that are responsible for the emission of the alpha-particle. But this knowledge contains the uncertainty which is brought about by the interaction between the nucleus and the rest of the world. If we wanted to know why the alpha-particle was emitted at that particular time, we would have to know the microscopic structure of the whole world including ourselves, and that is impossible" (*Physics and Philosophy* [London: Penguin Books, 1990], 78–79).
10 Even though the philosophical context is completely different, this Buddhist argument is rather similar to the following words of Hermes Trismegistus: "All things are struck by destruction. For without destruction there can be no generation. The things that are born need to arise from the things that are destroyed, and those that are born must be destroyed so that generation can continue. . . . It is impossible for the same things to be born again; and how can something that is not what it was before be real?"
11 Nagarjuna, *Fundamental Treatise on the Middle Way (Mulamadhyamakarika),* in *Wisdom: Two Buddhist Commentaries* (Paris: Éditions Padmakara, 1993).
12 For a thoughtful introduction to the teachings of Prajnapatamita, see *Profound Wisdom of the Heart Sutra and Other Teachings,* Bokar Rinpoche (Clearpoint Press, 1994).

CHAPTER 10. THE VIRTUAL FRONTIER

1 See Jean-Pierre Changeux, *Neuronal Man* (New York: Pantheon, 1985).

2 This explanation has been presented in terms of the "relative truth," i.e., the illusory world of phenomena such as they appear to us. The Dalai Lama does not suppose that cause and effect exist inherently, and he is not casting doubt over what has already been explained. The refutation in chapter 9 explains the impossibility of going from a cause to an effect if both are considered to be inherently existing entities.

3 Francisco J. Varela, ed., *Sleeping, Dreaming and Dying: An Exploration of Consciousness with the Dalai Lama* (Boston: Wisdom Publications, 1997), 119–20.

4 This is the "embodied cognition" concept. See F. Varela, E. Thomson, and E. Rosch, *The Embodied Mind* (Boston: MIT Press, 1991).

5 As Alan Wallace points out: "As soon as one begins to understand that subjective and objective, mental and physical phenomena are *relational* instead of *substantive,* the causal interactions between mind and matter become no more problematic than such interaction among mental phenomena and among physical phenomena. . . . Since the mind alone perceives both mental and physical events, as well as the relations between them, introspection should naturally play a vital role in determining such causal interactions."

6 But here, too, we mustn't confuse ultimate truth, which states that neither consciousness nor exterior phenomena exist inherently, and conventional or relative truth, which says that there is a *qualitative* difference between the animate and the inanimate. Even in a dream, a rock is different from a thinking being. However, this difference doesn't come from any basic duality, which is conceivable only if phenomena are taken to be real.

7 In Dilgo Khyentse Rinpoche, *The Heart Treasure of the Enlightened Ones,* translated by the Padmakara Translation Group (Boston: Shambhala, 1992).

8 Erwin Shrödinger, *Mind and Matter* (Cambridge, England: Cambridge University Press, 1958).

9 Not all biologists accept this theory. It has been criticized by, among others, Richard Dawkins.

10 See M. J. Meany et al., "Early Environmental Regulation of Forebrain Glucocorticoid Receptor Gene Expression: Implications for Adrenocortical Responses to Stress," in *Developmental Neuroscience* 18 (1996): 49–72.

11 Francisco Varela, personal communication.

12 Raymond Moody, *Life After Life* (New York: Bantam Books, 1977); Michael Sabom, *Recollections of Death* (New York: Harper & Row, 1982); and Kenneth Ring, *Heading Towards Omega* (Quill Morrow, 1984). See the critiques of these accounts in Varela, *Sleeping, Dreaming and Dying,* and Sogyal Rinpoche, *The Tibetan Book of Living and Dying* (San Francisco: HarperCollins, 1992).

13 See L. D. Gupta, N. R. Sharma, and T. C. Mathur (the three worthies sent by Gandhi), *An Inquiry into the Case of Shanti Devi* (Delhi: International Aryan League, 1936), as well as the article by Patrice Van Eersel in *Clés* 22 (summer 1999), which we have summarized here.

14 Ian Stevenson, *Twenty Cases Suggestive of Reincarnation* (Charlottesville: University of Virginia Press, 1974).
15 J.-F. Revel and M. Ricard, *The Monk and the Philosopher* (New York: Schocken Books, 2000). For the life of Khyentse Rinpoche, see *Journey to Enlightenment: The Life and World of Khyentse Rinpoche,* Matthieu Ricard (New York: Aperture, 1996).
16 Varela, ed., *Sleeping, Dreaming and Dying,* 216–17.

CHAPTER 11. ROBOTS THAT THINK THEY CAN THINK?

1 See the book by Marvin Minsky, an artificial intelligence expert, *Society of Mind* (New York: Simon & Schuster, 1987).
2 According to Francisco Varela (personal communication), if the consciousness is considered to be inseparable from the dynamic base that conditions it (the body and experiences), the distinction between hardware and software breaks down.
3 See the interview with Francisco Varela in *La Recherche* 308 (April 1998): 109.
4 Francisco Varela, personal communication.
5 Alan Turing, "Computing Machinery and Intelligence," in *Mind* 59 (1950): 433–60.
6 John R. Searle, "Minds, Brains and Programs," in *The Behavioral and Brain Sciences,* vol. 3 (New York: Cambridge University Press, 1980).
7 R. A. Brooks, "Intelligence without Reason," in *Proceedings of the 1991 International Joint Conference on Artificial Intelligence* (1991), 569–95; R. A. Brooks, "Intelligence without Representation," *Artificial Intelligence Journal* 47 (1991): 189–160; R. A. Brooks et al., "Alternative Essences of Intelligence," in *Proceedings of the American Association of Artificial Intelligence,* 1998.
8 Plants move to follow sunlight or capture their prey, but these are unconscious movements. For Buddhists, too, plants are not conscious.
9 See Varela, *La Recherche.*
10 Daniel Dennett, *Consciousness Explained* (Boston: Little, Brown, 1991), 21–22.
11 Alan Wallace, personal communication.
12 See Luc Steels, "The Artificial Life Roots of Artificial Intelligence," *Artificial Life Journal* 1, no. 1 (Cambridge, Mass.: MIT Press, 1994); also "The Homo Cyber Sapiens, the Robot Homonidus Intelligens, and the Artificial Life Approach to Artificial Intelligence," Burda Symposium on Brain-Computer Interfaces, Munich, 1995.
13 Varela, *La Recherche.*
14 David Potter in *An East-West Dialogue: The Dalai Lama and Participants in the Harvard Mind Science Symposium,* edited by Daniel Goleman and Robert A. F. Thurman (Boston: Wisdom Publications, 1991).
15 See Stevan Harnad, "Consciousness: An Afterthought," in *Cognition and Brain Theory,* chap. 5, 1982, 29–47.
16 See B. Alan Wallace, *The Taboo of Subjectivity: Toward a New Science of Consciousness* (New York: Oxford University Press, 2000), 29, 138–39; and Paul M. Churchland, *Matter and Consciousness: A Contemporary Introduction to the Philosophy of Mind,* revised ed. (Cambridge, Mass.: MIT Press, 1990).

17 Shantideva, *Bodhicaryvatara: The Way of the Bodhisattva,* vol. 10 (Boston: Shambhala, 1997), 55.
18 Francisco J. Varela, ed., *Sleeping, Dreaming and Dying: An Exploration of Consciousness with the Dalai Lama* (Boston: Wisdom Publications, 1997).
19 Ibid.

CHAPTER 12. THE GRAMMAR OF THE UNIVERSE

1 In a personal communication, Michel Bitbol has remarked that Kant's "objective reality" is synonymous with Buddhist "relative, or conventional reality." For Kant, who has supplied a large part of the philosophy of science's conceptual tools, intrinsic existence is "the thing in itself," which is unknowable, while objective existence consists in sequences of phenomena that are linked one to the other by categories of pure understanding: the principle of permanence (associated with the category of substance), the principle of consecution according to a rule (associated with the category of causality), etc.
2 Werner Heisenberg, *Philosophical Problems of Quantum Physics* (Woodbridge, Conn.: Ox Bow Press, 1979), 23.
3 Eugene P. Wigner, "The Unreasonable Effectiveness of Mathematics," *Communications on Pure and Applied Mathematics* 13 (1960): 1–14.
4 The difference between flat space and curved space can be explained by comparing a flat space with a plane. This is only an analogy, given that space has three dimensions and a plane only two, but it will be good enough to guide our intuition. At school, we all learned that a straight line can have only one parallel line passing through a given point, and that the angles of a triangle on a plane add up to 180 degrees. This is Euclidean geometry. Let's now take a curved surface. It can be convex, like the surface of a sphere, or else concave, like a saddle. On a convex surface, such as that of the Earth, the lines of longitude that look parallel at the equator converge at the poles. A straight line cannot have any parallel lines on the surface of a sphere. What is more, the sum of the angles of a triangle is greater than 180 degrees. On the other hand, on a concave surface a straight line can have a large number of parallel lines passing through a given point (a parallel line being defined as a line that never touches the straight line) and the sum of the angles of a triangle is less than 180 degrees.
5 For more details about fractal objects, see Trinh Xuan Thuan, *Chaos and Harmony* (New York: Oxford University Press, 2001), chap.3.
6 René Descartes, *Meditations on First Philosophy* (New York: Cambridge University Press, 1986).
7 Roger Penrose, *The Emperor's New Mind* (New York: Oxford University Press, 1989), 95.
8 Cited in Jacques Hadamard, *The Psychology of Invention in the Mathematical Field.* (Princeton: Princeton University Press, 1945), 13.
9 Roger Penrose, *The Emperor's New Mind,* pp. 97, 428.
10 Prime numbers are those that can be divided only by themselves and 1, for

example 1, 2, 3, 5, 7, 11, 13, etc. Other prodigies show an astonishing ability to memorize figures, such as the Japanese man who could recite *pi* to forty thousand decimal places.

11 For a synthesis, see "La biologie des maths," in *Science et Vie* no. 984 (September 1999): 46.

CHAPTER 13. REASON AND CONTEMPLATION

1 Gödel's proof is based on the concept of self-referring propositions, i.e., ones that describe themselves. The ancients were already familiar with the logical paradoxes that result from self-referring propositions. Take, for example, the statement "This sentence is false." If it's true, then it's false; but if it's false, it's true. Or the statement "I am a liar." If I am a liar, then I'm telling the truth; if I'm telling the truth, then I'm a liar. This completely floors logic. Similarly, the following problem posed by Bertrand Russell has no answer: "An inhabitant of Seville is shaved by the Barber of Seville if, and only if, he doesn't shave himself. So, does the Barber of Seville shave himself?" If he does shave himself, he can't be shaved by the Barber of Seville, so he can't shave himself. But if he doesn't shave himself, then he must be shaved by the Barber of Seville, and so he shaves himself. See E. Nagel and J. R. Newman, *Gödel's Proof* (New York: New York University Press, 1958).

2 Albert Einstein and Leopold Infeld, *The Evolution of Physics* (New York: Simon & Schuster, 1938), 31.

3 B. Alan Wallace, *Choosing Reality* (Ithaca, N.Y.: Snow Lion, 1996), 26–27.

4 Thomas Kuhn, *The Structure of Scientific Revolutions* (Chicago: University of Chicago Press, 1962).

5 Cited in Werner Heisenberg, *Physics and Beyond: Encounters and Conversations* (New York: Harper & Row, 1971), 63.

6 Thomas Kuhn, *The Structure of Scientific Revolutions* (Chicago: University of Chicago Press, 1962).

7 Norwood Russell Hanson, *Patterns of Discovery: An Inquiry into the Conceptual Foundations of Science* (Cambridge, England: Cambridge University Press, 1968), 30.

8 A. N. Whitehead, *Dialogues of Alfred North Whitehead, as Recorded by Lucien Price* (New York: New American Library, 1956), 109, quoted in Wallace, *Choosing Reality,* 11.

9 See Trinh Xuan Thuan, *The Secret Melody* (New York: Oxford University Press, 1995).

10 See B. Alan Wallace, *The Taboo of Subjectivity: Toward a New Science of Consciousness* (New York: Oxford University Press, 2000).

11 E. Rodriguez, N. George, J. P. Lachaux, J. Martinerie, Francisco J. Varela, "Perception's Shadow: Long-distance synchronization in the human brain," *Nature* 397 (1999): 340–43.

12 See Paul Ekman, Richard Davidson, Matthieu Ricard, and B. Alan Wallace, "Buddhist and Psychological Perspectives on Emotions and Well-Being" (in press).

CHAPTER 15. FROM MEDITATION TO ACTION

1 Shantideva, *Bodhicaryvatara: The Way of the Bodhisattva,* vol. 7, vv. 129–30 (Boston: Shambhala, 1997).
2 This theme is magnificently developed in the second chapter of Matthieu Ricard, *Journey to Enlightenment* (New York: Aperture, 1997).
3 General Jose Bonett, quoted in a 1998 BBC program.
4 The Peacemaker Order; consult their Web site, peacemaker@zpo.org.

THE MONK'S CONCLUSION

1 Niels Bohr, *Atomic Theory and the Description of Nature* (Woodbridge, Conn.: Ox Bow Press, 1987), 1.
2 Ibid., 18.
3 François Jacob, *Of Flies, Mice and Men* (Boston: Harvard University Press, 1998).
4 Michel Bitbol, *A Cure of Metaphysical Illusions: Kant, Quantum Mechanics and the Madhyamaka,* in B. A. Wallace, *Buddhism and Science* (in preparation).
5 Jacob, *Of Flies, Mice and Men.*
6 Matthieu Ricard, trans., *The Life of Shabkar: The Autobiography of a Tibetan Yogi* (Ithaca, N.Y.: Snow Lion, 2001).

THE SCIENTIST'S CONCLUSION

1 Steven Weinberg, "A Designer Universe?" *The New York Review of Books,* 21 October 1999, 46–48.
2 A study published in December 2000 has concluded that the size of the ozone hole is shrinking. This appears to be linked to the Montreal Protocol signed by many nations in 1987 to eliminate the chlorofluorocarbons (CFCs) used widely in aerosol sprays, air conditioners, and refrigerators, which are in large part responsible for the ozone depletion. This is a shining example of humanity at its best, working together to solve a global problem, the future of life on Earth. However, we still must exercise great vigilance. The hole is still huge, and at the present rate of shrinkage the hole will not disappear completely before 2060 at the earliest.
3 Quoted by Thinley Norbu in, "Welcoming Flowers," from *Across the Cleansed Threshold of Hope: An Answer to Pope's Criticism of Buddhism* (New York: Jewel Publishing House, 1997).

SCIENTIFIC GLOSSARY

Accelerator: A machine using electric fields to accelerate electrically charged particles (such as electrons, protons, or their antiparticles) to give them a high energy. Because linear accelerators would require impractical lengths to reach high energies, most accelerators are circular. They use magnets to bend the path of particles, which pick up more energy at each new turn around the loop.

Animism: Philosophy attributing a soul to natural phenomena and objects.

Atomic nucleus: The most massive part of an atom, composed of protons and neutrons, around which electrons orbit. The nucleus is 100,000 times smaller than an entire atom; therefore, matter is made almost entirely of void.

Anthropic principle: The notion that the universe was tuned with extreme precision to allow for life and consciousness to emerge. The word comes from the Greek "anthropos," which means "man."

Antimatter: Matter composed of antiparticles, such as antiprotons, antielectrons (or positrons), and antineutrons. Antiparticles have exactly the same properties as their corresponding particles, except for the electrical charge, which is of opposite sign.

Antiparticle: A constituent of matter with the same properties as its matching particle, with the exception of the electrical charge, which is reversed.

Atom: The smallest particle of an element that still retains the properties of that element.

Behaviorism: Psychological doctrine emphasizing behavior as object of study and observation as method; it excludes anything that is not directly observable, such as thought.

Big Bang: Cosmological theory, according to which an extremely hot and dense universe would have been created in a huge explosion occurring everywhere in space about 15 billion years ago.

Big Crunch: The opposite of the Big Bang. The hypothetical final stage of the universe collapsing in on itself under the influence of its own gravity. No one knows if the universe contains enough matter for gravity to eventually stop and reverse the current expansion.

Black hole: Celestial object collapsed on itself, usually resulting from the death of a massive star. Its gravity is so strong that neither matter nor light can escape.

Butterfly effect: A phenomenon such that a very small change in the initial state of a dynamical system can dramatically alter its subsequent evolution.

Chaos: Property characterizing a dynamical system whose behavior depends very sensitively on the initial conditions.

Complementarity principle: Principle stated by the Danish physicist Niels Bohr, which asserts that matter and radiation can behave both as waves and particles, these two descriptions of Nature being complementary.

Cyclic universe: It goes through a succession of Big Bangs and Big Crunches and has neither beginning nor end.

Dark matter: Matter of unknown nature that does not emit any radiation. It may make up between 90 and 98 percent of the mass of the universe. Its existence is deduced from the gravitational influence it exerts on the motion of stars and galaxies.

Demiurge: Supreme Being who, according to Plato, exists in space and time and fashions the material world after plans in the world of Ideals ruled by the Good, an eternal and immutable Being existing outside time and space.

Determinism: Philosophical doctrine according to which there exist cause-and-effect relationships between physical phenomena, making it possible to predict their behavior if one knows the initial conditions.

Ecosphere: The soil, water, and air environment in which living beings evolve on Earth.

Electron: The least massive stable elementary particle. Electrons carry a negative charge, and, together with protons and neutrons, are constituents of atoms.

Electromagnetic force: It is responsible for the property that particles with opposite charges attract each other, while those with like charges repel each other. It binds together atoms and molecules.

Electromagnetic spectrum: The set of all types of radiation, from radio waves (the least energetic) to gamma rays (the most energetic).

Emergent property: Refers to a property of a complex system that cannot be deduced or explained in terms of the properties of its constituents. In other words, the whole is greater than the sum of its parts.

Fossil radiation: Radio radiation that bathes the entire universe, dating back from the time the universe was only 300,000 years old. It is the remnant heat from the Big Bang. Because of the expansion of the universe, it has cooled down considerably, and its present temperature (that of intergalactic space) is only -270 degrees C.

Fractal object: Object whose spatial dimension is not an integer. It may also refer to an object whose dimension is an integer, but which displays patterns that repeat themselves ad infinitum regardless of magnification.

Galaxy: Large system containing on average several hundred billion stars bound together by gravity. It is a fundamental building block of the vast structures in the universe.

General relativity: A theory developed by Einstein in 1915, which relates accelerated motions to gravity and the geometry of space-time.

Gravitational force: Force responsible for the attraction of one material object toward another; it is proportional to the product of the masses of both objects, and inversely proportional to the square of the distance separating them.

Holism: Philosophical doctrine in opposition to reductionism. Whereas reductionism asserts that the whole can be decomposed and analyzed in terms of its constituents considered as fundamental, holism professes, on the contrary, that the whole is fundamental and cannot be reduced to its components, as the whole is sometimes greater than the sum of the components.

Idealism: Philosophical doctrine in which any phenomenon external to man is subordinated to thought.

Ideals, world of (or world of Forms): According to Plato, the world of the senses is changing, ephemeral, and illusory; it is only a pale reflection of the world of Ideals, which is eternal, immutable, and genuine.

Incompleteness theorem: Theorem discovered by the Austrian-American mathematician Kurt Gödel; it states that any arithmetic system contains undecidable propositions that can be neither proved nor disproved by means of the axioms contained within that system.

Initial conditions: The state of a dynamical system at the start of its evolution.

Light-year: The distance covered by light (which travels at 300,000 km/s) in one year; it is equal to 9,460 billion kilometers, or 5,910 billion miles.

Linear system: System in which changes in the initial conditions lead to proportional changes in the final state.

Mach's principle: According to Austrian physicist Ernst Mach, the mass of an object is determined by the distribution of all the matter in the universe through a mysterious interaction.

Materialism: Philosophical doctrine professing that nothing exists besides matter, and that the mind itself is wholly material.

Meson: A particle composed of one quark and one antiquark.

Molecule: Combination of one or more atoms, bound by the electromagnetic force.

Neutrino: Elementary particle without electrical charge and with a mass that is either zero or extremely small. It interacts very weakly with ordinary matter.

Neutron: Electrically neutral particle composed of 3 quarks; together with protons, it is a constituent of atomic nuclei. A free neutron has a lifetime of about fifteen minutes. It decays into a proton, an electron, and an anitineutrino. When inside an atomic nucleus, however, it does not decay, being then as stable as a proton.

Nonlinear system: System in which changes in initial conditions do not produce proportional changes in the final state.

Occam's razor: The notion that a simple explanation for a phenomenon is more likely to be true than a more complicated one. The term "razor" refers to "shaving off," that is to say, eliminating any superfluous hypothesis. Occam's razor is attractive because it satisfies our sense of beauty and elegance.

Parallel universes: Universes existing simultaneously but completely disconnected from our own and, therefore, not accessible to observation. Quantum mechanics as well as certain theories of the Big Bang predict the existence of such parallel universes.

Periodic table: List of chemical elements in increasing order of atomic numbers, grouped in columns according to their reactive properties. It was discovered by the Russian chemist Dmitri Mendeleyev.

Photon: Particle of light. With no mass or charge, it travels at 300,000 km/s.

Planck's length: Equal to 10^{-33} cm, it is the dimension at which space becomes a quantum foam and known physics break down. It is also the length of superstrings.

Planck's time: Equal to 10^{-43} s, it is the shortest time interval that can exist. Known physics break down for time intervals smaller than Planck's time.

Planet: A body orbiting around a star in a solar system. Unlike stars, planets do not possess their own internal energy source such as nuclear energy. The radiation they emit is due nearly entirely to reflection of the light from the star.

Proton: Particle with a positive charge, composed of 3 quarks. Together with the neutron, it is a constituent of atomic nuclei.

Quantum fuzziness: See "Uncertainty Principle."

Quantum gravity: Theory (yet to be developed) that would unify the two pillars of modern physics—quantum mechanics and general relativity. Such a theory would enable us to go beyond Planck's wall, which currently constitutes a barrier to our knowledge.

Quantum mechanics: A branch of physics that describes the structure and behavior of atoms and their interactions with light in terms of probabilities. In this theory, the energy, spin, and other quantities are quantized, that is to say, they can vary only in discrete amounts that are multiples of a unit value. The phenomena predicted by quantum mechanics include quantum fuzziness, wave-particle duality and virtual particles.

Quantum vacuum: Space filled with virtual particles and antiparticles that appear and disappear in exceedingly short life and death cycles; their existence is related to the energy fuzziness resulting from the uncertainty principle.

Quark: Hypothetical particle supposed to be the most fundamental constituent of matter. It has a fractional electrical charge that is either positive or negative, equal to $^1/_3$ or $^2/_3$ of the charge of an electron. No quark has ever been seen in a free state. Quarks combine in groups of three, bound together by the strong nuclear force, to form a proton or a neutron. Six different kinds of quarks are known: up, down, strange, charm, bottom, and top, each coming in three colors (yellow, red, and blue).

Quasar: Celestial object that is among the most distant and brightest objects in the universe. Their enormous energy comes supposedly from a supermassive black hole with a billion solar masses, devouring the stars of the associated galaxy.

Reductionism: Method for studying a physical system by decomposing it into its most elementary constituents considered as fundamental.

Special relativity: A theory developed by Einstein in 1905, dealing with relative motions; it establishes an intimate connection between time and space, which are no longer absolute and universal but depend on the motion of the observer. It also establishes the equivalence between matter and energy.

Steady-state theory: Cosmological theory that maintains that the universe is unchanging in space and time. To compensate for the empty space created by the expansion of the universe, the theory postulates a continuous creation of matter.

Strong nuclear force: It binds quarks to form protons and neutrons, and protons and neutrons themselves to form atoms.

Superstring theory: A theory based on the notion that elementary particles of matter are not points, but vibrations of infinitesimally small bits of string with length equal to Planck's length.

Supernova: Explosive death of a massive star (with more than 1.4 times the mass of the Sun) after it has exhausted its nuclear fuel.

Tachyon: Hypothetical particle traveling faster than light.

Turing test: Test proposed by the British mathematician Alan Turing for determining whether a machine is or is not endowed with intelligence.

Uncertainty principle: Discovered by the German physicist Werner Heisenberg, it states that the velocity and position of a particle cannot be measured simultaneously with arbitrary precision, no matter how sophisticated our measuring instruments. It is sometimes referred to as quantum fuzziness. The uncertainty principle also applies to the energy and lifetime of a particle. The fuzziness of the energy allows for the existence of virtual particles and antiparticles.

Virtual particle: Particle created as a pair with its matching antiparticle (the total electrical charge must remain zero) by borrowing energy from an adjacent region of space. In accordance with the uncertainty principle, the amount of energy borrowed must be returned very quickly, so the virtual particles disappear in a very short time and cannot be detected directly by our instruments. Virtual particles can materialize into real particles when energy is being fed in, as was the case in the first moments of the universe.

Vitalism: Doctrine according to which biological systems cannot be reduced to collections of molecules and their interactions but possess a life principle distinct from both the soul and the organism.

Wave-particle duality: The property of light or matter to behave sometimes as waves, sometimes as particles.

Weak nuclear force: It is responsible for radioactivity. It transforms one particle into another. For instance, it is responsible for the decay of a free neutron into a proton in about fifteen minutes.

BUDDHIST GLOSSARY

Absolute truth: The ultimate nature of the mind and the true status of all phenomena; the state beyond all conceptual constructs that can be known only by primordial wisdom and in a manner that transcends duality. The way things are from the point of view of realized beings.

Actions: Actions resulting in the experience of happiness for others and defined as *positive* or virtuous; actions that give rise to suffering for others and oneself and are described as *negative* or nonvirtuous. Every action, whether physical, mental, or verbal, is like a seed leading to a result that will be experienced in this life or in a future life.

Afflictive mental factors, or negative emotions (sanskrit: *klesha*): All mental events born from ego-clinging that disturb the mind and obscure it. The five principal afflictive mental factors, which are sometimes called "mental poisons," are attachment, hatred, ignorance, envy, and pride. They are the main causes of both immediate and long-term sufferings.

Aggregates, five, (sanskrit: *skandha*), *lit.* "heaps," "aggregates," or "events": The five aggregates are the component elements of form, feeling, perception, conditioning factors, and consciousness. They are the elements into which the person may be analyzed without residue. When they appear together, the illusion of "self" is produced in the ignorant mind.

Appearances: The world of outer phenomena. Although these phenomena seem to have a true reality, their ultimate nature is emptiness. The gradual transformation of our way to perceive and understand these phenomena corresponds to the various levels of the path to enlightenment.

Awareness, pure: The nondual ultimate nature of mind, which is totally free from delusion.

Bardo: Tibetan word meaning "intermediary state." This term most often refers to the state between death and subsequent rebirth. In fact, human experience encompasses six types of bardo: the bardo of the present life, the bardo of meditation, the bardo of dream, the bardo of dying, the luminous bardo of ultimate reality, and the bardo of becoming. The first three bardos unfold in the course of life. The second three refer to the death and rebirth process that terminates at conception at the beginning of the subsequent existence.

Bodhisattva: One who through compassion strives to attain the full enlightenment Buddhahood for the sake of all beings.

Buddha: One who has eliminated the two veils—the veils of emotional obscurations and the cognitive obscuration, which is the dualistic conceptual thinking, which prevents omniscience—and who has developed the two wisdoms, the wisdom that knows the ultimate nature of the mind and phenomena, and the wisdom that knows the multiplicity of these phenomena.

Buddha nature: It is not an "entity" but the ultimate nature of mind, free from the veils of ignorance. Every sentient being has the potential to actualize this Buddha nature by attaining perfect knowledge of the nature of mind. It is in a way the "primordial goodness" of sentient beings.

Clinging, grasping, attachment: Its two main aspects are clinging to the true reality of the ego, and clinging to the reality of outer phenomena.

Compassion: The wish to free all beings from suffering and the causes of suffering (negative actions and ignorance). It is complementary with *altruistic love* (the wish that all beings may find happiness and the causes of happiness), *sympathetic joy* (which rejoices of others' qualities), and *equanimity;* it extends the three former attitudes to all beings, whether friends, strangers, or enemies.

Consciousness: Buddhism distinguishes various levels of consciousness: gross, subtle, and extremely subtle. The first one corresponds to the activity of the brain. The second one is what we intuitively call "consciousness," which is, among other things, the faculty of consciousness to know itself, investigate its own nature, and exert free will. The third and most essential one is called the "fundamental luminosity of mind."

Dharma: This Sanskrit term is the normal word used to indicate the doctrine of the Buddha. The Dharma of transmission refers to the corpus of verbal teachings, whether oral or written. The Dharma of realization refers to the spiritual qualities resulting from practicing these teachings.

Duality, dualistic perception: The ordinary perception of unenlightened beings. The apprehension of phenomena in terms of subject (consciousness) and object (mental images and the outer world), and the belief in their true existence.

Ego, "I": Despite the fact that we are a ceaselessly transforming stream, interdependent with other beings and the whole world, we imagine that there exists in us an unchanging entity that characterizes us and that we must protect and please. A thorough analysis of this ego reveals that it is but a fictitious mental construct.

Emptiness: The ultimate nature of phenomena, namely their lack of inherent existence. The ultimate understanding of emptiness goes together with the spontaneous arising of boundless compassion for sentient beings.

Enlightenment: Synonymous with Buddhahood. The ultimate accomplishment of spiritual training. Consummate inner wisdom united with infinite compassion. A perfect understanding of the nature of mind and of phenomena, that is, their relative mode of existence (the way they appear) and their ultimate nature (the way they are). Such understanding is the fundamental antidote to ignorance and therefore to suffering.

Existence, true, intrinsic, or reality: A property attributed to phenomena, suggesting that they could be independent objects, existing in themselves, and having local properties that belong to them intrinsically.

Idealism: Set of ideas stating that the world of phenomena is simply a projection of the mind.

Ignorance: An erroneous way to conceive of beings and things, which consists in attributing to them an existence that is real, independent, solid, and intrinsic.

Illusion: All ordinary perceptions, deformed by ignorance.

Impermanence: It has two aspects: gross impermanence is applied to visible changes; subtle impermanence reflects the fact that nothing can remain identical to itself, even for the shortest conceivable instant.

Interdependence or "dependent origination": A fundamental element of Buddhist teaching, according to which phenomena are understood not as discretely existent entities but as the coincidence of interdependent conditions.

Karma: A Sanskrit word meaning "action," usually translated by "causality of actions." According to the Buddha's teachings, beings' destinies, joys, sufferings, and perceptions of the universe are due neither to chance nor to the will of some all-powerful entity. They are the result of previous actions. In the same way, beings' futures are determined by the positive or negative quality of their current actions. Distinction is made between collective karma, which defines our general perception of the world, and individual karma, which determines our personal experiences.

Liberation: To be free from suffering and the cycle of existences. This is not yet the attainment of full Buddhahood.

Logic: A correct approach to knowledge (*pramana* in Sanskrit, *tsema* in Tibetan). Distinction is made between valid "conventional" knowledge and valid "absolute" knowledge. The former teaches us about the appearance of things, while the latter allows us to perceive the ultimate nature of phenomena. Both are valid in their respective registers. Their field covers everything that can be directly perceived or else deduced by inference and what can be concluded on the basis of reliable testimony.

Meditation: A process of familiarization with a new perception of phenomena. Distinction is made between analytical meditation and contemplative meditation. The object of the former could be a point to be studied (for instance, the notion of impermanence) or else a quality that we wish to develop (such as love and compassion). The latter allows us to recognize the ultimate nature of the mind and to remain inside that nature, which lies beyond conceptual thought.

Middle way (*madhyamika*): Buddhism's most elevated form of philosophy, so called because it avoids the two extremes of *nihilism* and of belief in the reality of phenomena (*eternalism* or *materialism*).

Mind (*see also consciousness*): In Buddhist terms, the ordinary condition of the mind is characterized by ignorance and illusion. A succession of conscious instants gives it an appearance of continuity. In its absolute form, the mind is defined by three characteristics: emptiness, clarity (ability to know all things), and spontaneous compassion.

Nirvana: "Beyond suffering," expresses several levels of enlightenment, depending on whether our viewpoint is from the Small Vehicle or the Great Vehicle.

Path: The spiritual training that allows us to free ourselves from the cycle of existences, then reach the state of Buddhahood.

Phenomena: What appears to the mind, through sensory perceptions and mental events.

Realism, reification: *See Existence.*

Rebirth, reincarnation: The successive states that are experienced by the flow of consciousness, and which are punctuated by death, bardo (q.v.), and birth.

Relative truth, *Lit.* "all-concealing truth." This refers to phenomena in the ordinary sense, which, on the level of ordinary experience, are perceived as real and separate from the mind, and which thus conceal their true nature.

Samsara: The wheel or round of existence; the state of being unenlightened, in which the mind, enslaved by the three poisons of desire, anger, and ignorance, evolves uncontrolled from one state to another, passing through an endless stream of psycho-physical experiences, all of which are characterized by suffering. It is only when one has realized the empty nature of phenomena and dispelled all mental obscurations that one can free oneself from samsara.

Suffering: The first of the "four noble truths," which are (1) the truth of suffering, which must be seen as being omnipresent in the cycle of conditioned existences, (2) the truth of the origin of suffering—the negative emotions that we must eliminate, (3) the truth of the path (spiritual training) that we must take in order to reach liberation, and (4) the truth of the cessation of suffering, the fruit of training, or the state of Buddhahood.

Sutra: The words of Buddha Shakyamuni, which were transcribed by his disciples.

Thoughts, discursive: A normal linking together of thoughts conditioned by relative reality.

Universe, cyclic: A universe governed by cycles, each one having four phases. The first corresponds to the formation of the universe, the second to its evolution, and the third to its destruction. The fourth corresponds to a period of vacuity that separates two universes. The continuity between two universes is guaranteed by potential manifestations, called "space particles." These cycles are successive but nonrepetitive.

View, meditation and action: The vision of emptiness must be integrated into our mind via meditation, which must in turn be expressed in altruistic actions.

Wisdom: (1) The ability to discern correctly, (2) the understanding of emptiness, and (3) the primordial and nondual knowing aspect of the nature of the mind.

ACKNOWLEDGMENTS

We would like to thank everyone who inspired this book: Maria-Angels Vilana, who invited us to the Summer University in Andorra, and who was thus the initial cause of our meeting and the origin of our dialogue, and our editors Nicole Lattès and Claude Durand, who encouraged us to write this book and who constantly made us go over our work in order to make it more profitable for our readers.

We also must express our gratitude to Christian Bruyat (who transcribed the recordings of our conversations and edited them) and Carisse Busquet and Dominique G. Marchal (who reread and improved the various drafts of our manuscript). Along with Gérard Godet and Yahne Le Toumelin, they were our attentive correctors. For this English edition we are indebted to Ian Monk, who carried out the translation work with unusual speed and competence, and put up with our multiple suggestions and queries, and to our editor Emily Loose, who pursued this project with unflagging enthusiasm and warmth and made a major contribution to bringing this book to its final shape.

We are also particularly grateful to our friends and family, who were good enough to show interest in our project and to give us vital suggestions and criticism concerning its contents: Michel Bitbol, Francisco Varela, Allan Wallace, Jean-François Revel, Wulstan Fletcher, Abel Gerschenfeld, and Miguel Benasayag. Nguyen Tan Nam's computer assistance was also invaluable.

Finally, we would like to thank all those who answered our questions, who helped us to find the information we required, who participated in the making and the promotion of this book, or who otherwise encouraged us: Brian Greene, Laurent Nottale, Catherine Bourgey, Susanna Lea, Vivian Kurz, Mark Tracy, Françoise Grandgirard, Jeanne Gruson, Hélène Boullet, Laurence and Bruno Bardèche, Jean Staune, and René Dubois.

The scientist's special thanks go to Georges Alecian, Chantal Balkowski, and François Hammer for their hospitality at the department of extragalactic astronomy and cosmology at the Paris-Meudon observatory, as well as Bernard Fort and Bruno Guiderdoni for having welcomed him to the Institute of Astrophysics in Paris during his sabbatical year.

INDEX